We Are Only Human

By

James J. Weitzman

Contents

Dedication

This book is dedicated to my entire family, including our golden retrievers, without whom I could not have written it. Their encouragement, humor, and enthusiasm have given me the motivation to share what I believe it means to be human.

Acknowledgements

There are so many elements that I have collected and shared in my life that acknowledge the content in *We Are Only Human.* That would take years to fully establish. The first clarification to be stated is that I do not profess to be any degree of an authority on the matter of human nature or human behavior. I make observations. This has enabled me to be in tune with reality. My opportunity to have taught biology, human anatomy and physiology, evolution, and biological anthropology has aided me in developing the knowledge for this book. But perhaps most important are my years spent living this life. In doing so, I have lived a thousand lives and experienced pain and sorrow, as so many others have. This has granted me some insight into what it means to be human. With that being said, I feel gratefully blessed. With clear vision, I can recognize the challenges that life demands. I can acknowledge that I am very fortunate to have lived a fantastic life.

Much of what I have learned, and more importantly experienced, has shown me that we are all more alike than different. I believe this is what makes *We Are Only Human* very relatable to most people. Many of us have ventured through the same types of occurrences at least once or more in our lives. Therefore, my acknowledgment in this book is to all of you. Thank you for being human.

I give the utmost thanks to our ancestors for being who they were, to close friends who know me so well, and most especially to my family. If not for each of you, I could not write this book. You continue to be my inspiration, encouragement, and undeniable satisfaction. Without my true personal family stories told in this book, many others may not relate to their own human history. All of you make us human and yet vulnerable. However, none of this would have come to fruition without your loving presence.

About the Author

James J. Weitzman is a biologist and former educator of life sciences, human anatomy and physiology, evolution, and biological anthropology. He lives in New Jersey with his wife, Barbara. He has two children and several golden retrievers. He is a master swimmer, loves all water activities, and is an avid physical training enthusiast. James also has two grandchildren and is a lifelong musician and composer. His prior book, *A Belief to Build a Life On*, is dedicated to his children.

Chapter 1: A New Beginning

The morning light is dim. The sun has not risen, and still the day promises to be hot with minimal moisture from above. The air is still. The horizon appears as a mosaic of color in the early eastern sky. The quietness roams across the plains of the African savannah. However, life is in abundance, yet unseen to the watchful eye of all prey. Sudden movements spark interest and piercing observations from waiting predators hidden from the cautious eyes of potential victims. Vigilance is at its highest peak, and the instincts for the survival of both prey and predators alike are alert and keen from their previous experiences and innate abilities.

Daylight in the open plains of the African savannah four to five million years ago is as intense with the competition for survival as it is today. The absence of light provided by the darkness of night serves little benefit during the daylight hours to all prey, regardless of his or her health, skills, or sharpened senses. The weak, the sick, and the old are the most disadvantaged in any hunt. But it is nature's design to select these individuals to be sacrificed for the health of the herd and the survival of the minority predator. It comes down to the survival of the fittest. The strong remain for the strength of the herd, and food is provided to a well balanced, stable, healthy ecosystem. Death in this setting appears cruel and harsh, but necessary for a continuance within this fine balance of nature.

Our human ancestors endured these times in the great plains of Africa millions of years ago. Early humans were much smaller in stature, quite timid, non aggressive, and survived by running and hiding from potential danger and threats rather than fighting the fierce beasts of the world. There were no egos to stand and fight or the shame of dishonor in fleeing. Instincts drove our ancestors to behave in a manner which would increase their chance of survival. Our emotions of pride and honor fortunately did not interfere with early humans making unwise decisions or choices which would jeopardize themselves or their small group of companions. We were driven by instinct or nature's demands.

However, some may disagree about their course of action early on. Our ancestors spent much of their time in the safety of trees rather than on the ground. I sometimes wonder if the first of our ancestors who ventured out of the trees and onto the savannah soil did so because one of their kin "*triple dog dared*" them to do it. If so, the rest is history. Nonetheless, a five million year period of evolution would change not only our anatomy and physiology, but also our behavior, intellect, and ability to willingly go against those innate forces embedded in our genes which aided in our survival years ago. Soon to arrive were the well designed, undoubtedly the most destructive force, but potentially the most impactful creatures to ever exist: mankind. We are what we are because we are only human.

The question of what it is to be human, at first glance, may appear to be easy to answer. However, when one looks deeper into the complexities of our species, we quickly see how multilayered we really are. Obvious answers to apparently simple questions are by no means a challenge. But to be human is totally different. To best describe what it is to be human may take a lifetime. Since we are human, we must therefore know what it is to be human, or do we?

If one could gaze back in time, we do have some understanding of our earliest ancestors by the artifacts they left behind. Physical anthropologists could compare and see distinct differences between those ancestors who existed three million years ago and those who had lived three thousand years ago. These ancestors lived an animal lifestyle. Physical evolution between groups of humanoids would be apparent based upon fossil evidence. So where in our relatively short timeline of existence had we progressed from early human like to human? In order to fully satisfy the question, we must set parameters for what it means to be human.

Every year in my biology classes, I would ask the question at the end of our topic in evolution: "*What does it mean to be human?*" Most answers from my sophomore students included physical traits that we currently possess. You will always hear: bipedal, opposable thumb, a large skull and brain, intelligence, good vision, change in the jaw and size of the teeth to consume meat and plant material, and the ability to manipulate objects with the hands and fingers. This would allow us the ability to fabricate and use tools and weapons and, of course, be smart enough to make and use fire.

I would then shake my head and ask my students to tell me something that is not so obvious. The previous answers are common knowledge to everyone who has had basic biology. However, it does not necessarily answer the question. There is a laundry list of other species that also have similar qualities. The answer to "what does it mean to be human" is actually a complex and forever changing subject. We need to examine human nature and human behavior to figure out who and what we really are.

Five million years ago in the plains of Africa, our earliest ancestor by the genus *Australopithecus,* as an adult, stood to be the height of a small child of today. They were very small in stature, walked upright on two legs, but were not an aggressive species, which is often depicted by the brute "caveman" in movies and stories. *Australopithecus* had more ape like features than those of modern humans. Consequently, early uneducated opinions of our evolution linked us to monkeys or apes.

First of all, monkeys and apes are quite different. For one, monkeys have tails, apes do not. Humans did not ascend from apes or monkeys. We cannot make a link or compare entirely different species with humans. Although I have witnessed kids running around like little monkeys and adults acting like jackasses. Not an evolutionary observation! But this would be a case of using behavioral analogies and not anatomical or physiological characteristics. However, the appearance of *Australopithecus* did have ape like features: small skull and brain, protruding face and jaw, and substantial hair for protection, to name only a few traits.

Evolution is entirely about *change*. It is the change in DNA over time. But most people view evolutionary change as physical change. There is more to change than simply what we can see with our eyes. And to answer the question about what it means to be human, we need to dig deep and search with unbiased, open minded vision. It is a complex question to answer for a complex organism such as us.

As time moved onward, our ancestors changed, adapted, and learned quickly. The advantage of the human brain lies in the data that it discovers. This information is mentally tested, applied for use, stored, recalled, and used again. This is the basis for determining what we have learned. Often, understanding of such data comes later. But early ancestors were able to make discoveries, remember, and use the information that was tested through trial and error. Eventually, humans began to quickly learn about the world around them and advance by maximizing the skills that they had learned. Knowing how to use something doesn't always mean that one understands it. Hopefully, these abilities will soon follow. If this occurs, then the opportunity for wisdom may enter.

Wisdom is the ultimate end, for wisdom implies being more than smart or skillful. True wisdom offers vision beyond our normal range of sight, and vision opens up all possibilities to all potential realities. Real dreams for potential opportunities are based upon such vision.

A new beginning can emerge daily when considering changes in the DNA of our early ancestors. This certainly would produce physiological adaptations that enabled them to have a better chance to survive and pass along these molecular variations to new generations. Evolution simply offers opportunities to outcompete one's competitors and continue the genetic makeup of the species. Any trait in a population with optimum qualities enhances one's chance for survival. However, good traits adapted by just luck don't determine success of survival. It's as if someone were to give you a lottery ticket. You may not win, but if you didn't have a ticket to begin with, then you wouldn't even have a remote chance of hitting the numbers. So, are you feeling lucky? I myself always feel extremely lucky until the lottery number is drawn, at which point I need to wait a full week to feel lucky again. It requires more than a little luck when it comes to evolution.

Chance mutations and being in the right place at the right time may lead to survival, but evolution plays the game of survival of the fittest, or rather, which group is the best *adapted.* But is there a part of evolution that doesn't play according to the rules of being most adapted? And what part of a species' continuance does a designed arrangement undertake? That would lead us into a whole new discussion. This may not be part of the scientific equation for evolutionary success, but when all accounts are examined for a specific species, such as us, it makes one wonder what other forces, other than the nature of evolution and chance, are at work. Fortunately, evolution does not play by luck alone.

Our ancestors underwent tremendous physiological changes over our five million year period. We could sit down and rationalize the variations of the traits, such as our ability to stand and walk upright on two legs. This allowed our mates to visually see one another within the African plains and, equally important, see danger approaching. The deer staring in the headlights routine began for us over five million years ago. Our increased skull has accounted for an increased brain mass. The human brain got larger. Early humans went from being herbivores to scavengers to finally hunters eating massive amounts of meat. Increase the intake of protein from meat and the brain gets larger.

The human brain today weighs about three pounds. Talk about size? Consider the Blue Whale, who has a six foot penis! Even though the Blue Whale's brain weighs approximately fifteen pounds, it doesn't mean it is smarter than you. Sorry men, whatever you're thinking, it's physically impossible. Don't attempt to get a bigger brain to increase the size of your manhood. I suppose there are some things that are equally rewarding other than just intelligence and the processing of information using a large brain. But you're born with what you have.

Over time, the face of humans became flattened vertically, as opposed to having a protruding jaw. The muscles of the jaw reduced in size, and the shape of the teeth also changed based upon what we were eating thousands of years ago. Changing from plant based food to meat requires different types of teeth. The amount of hair covering our bodies became less and less due to limited advantages. We developed the ability to sweat to cool our bodies because of an increase in mobility and shifting climates. And all of this took time. A great deal of time.

Humans migrated from place to place looking for food or more space. Movement produces heat, therefore the need for sweat glands. Those adaptations two to three million years ago were traits from the past. It continues today and hasn't changed much. Makes one wonder about our future.

We survived because of the right mutations at the proper time. Thousands of different species of life adapted and yet became extinct over a millennium. But is this the only reason why humans endured a very difficult struggle of survival? Or does this only define us as humans? Our history and survival as a species, by all accounts, pivots on a delicate fine line of the right adaptation at the right time under the proper circumstances. We have only been on earth for a short period of time, considering the magnitude of our planet's history. Was it a bit of luck or a blend of a destined plan for future survival? How are the numerous behavioral changes and risks of our past compared to today's behaviors for tomorrow's survival?

The small groupings of Australopithecines, sparsely dispersed throughout mostly Africa millions of years ago, were a meek, timid, frail species desperately trying to find their niche within the environment. Their survival required them to live together in a never ending search for food and constantly attempt to avoid danger by hiding or running away to avoid conflict and risk of death. Behavioral traits did not include ego, desire, type of power, or control as often exhibited today. It wasn't like we had politicians back then. Somehow today we lost sight of the value of a leader leading for the sake of their tribe and not for personal gains or interests, as do many now. Politicians don't *lead*; heroes do.

Our ancestors needed to be submissive to life and subordinated to nature's powers. These early humans did not know peace because they did not know war. There was most likely limited aggression against one another, only the desire to survive. Innate emotions of fear and feelings of uncertainty stimulated their sense to avoid danger and not to stand and demonstrate power against a threatening foe. They contained and displayed animal behavior. It is nature's way to balance, preserve, and maintain order of living organisms within a growing, changing ecosystem.

It was only later in our history when early humans learned the seeming advantages of desire, greed, and power over others for personal gains. However, these acts of behavior are often found to be temporary, unobtainable, or sometimes elusive without our ability to exercise them as originally intended. The emotional evolution of humans even today has not reached its potential or risen to its *hopeful level.* Nonetheless, it is a behavior that needs to be placed within the parameters of what it is to be currently human. We may no longer need to be submissive and roll over showing our bellies, but certainly humble to a true threat. However, being respectfully humble is a worthy goal that is often forgotten. There is a time and place for all things, while not always reacting to simple emotional influences.

Human nature quickly developed and grew into many different and unique aspects of behavior that classify our diverse range of traits. With the intuition to develop tools, the ability to recall information and experiences, and physiological advances of the nerve network of our right and left hemispheres of the brain, we have become multidimensional when compared to other life forms. Animal nature can be predictable under set conditions within an environment. Nonetheless, we are *still* animals, and the definition of being human has almost limitless characteristics. However, a new beginning had begun.

As predictable as some human behavior can be, our nature to implement these behaviors is forever evolving. Today we seem more influenced and fascinated by that which gives instant satisfaction or amusement. Perhaps it is a result of experienced feedback from specific social media platforms or when it comes to winning at various competitions. This is clearly demonstrated by online gaming obsessions. Who wouldn't get instantly hooked? I apologize for this analogy. I may be one of the few exceptions. Gaming serves no interest to me. The gaming industry plans on most people's addictive behavior for pleasure.

Take it a step further and look at the numerous reality TV programs that are now presented every day. Who would have expected a show like "Hillbilly Hand Fishing" to be a weekly program? I'm sorry, but I just don't get the reality TV stuff. People do know that this is not actual reality. At the very least, I hope. It never caught on in our family. Nonetheless, are we an outstanding species seeking trivial joy in our lives, or what? Hopefully Hillbilly Hand Fishing and other "*reality*" TV programs will diminish. On the other hand, I unfortunately doubt it. I don't understand the attraction. Perhaps it is *escapism* from true reality rather than mindless television attractions. I'm throwing you a bone here, people, hoping you are more realistic and find satisfaction in other insightful entertainment.

Overall, my desire is that most people want to understand the unknowns of our world rather than pursue momentary interests. Scientists draw pleasure in assuming they know life's mysteries. However, a *true* scientist would be delighted in recognizing the *failure* of their work. This would drive those individuals to research even further to find the truth to real questions.

People have a specific amount of knowledge about certain living organisms because of our ability to make intensive observations. Analyze the surrounding area. Draw sound conclusions! That should be enough, right? We become obsessed with the need to acquire knowledge. Our interests are like addictions, living in a realm of their own. However, humans play outside of the boundaries of common social or instinctual behavior, which drives us toward specific personal interests and pleasures. This broad range of characteristics places humans in a category unlike any other living organism. We're not just an ordinary animal among the thousands of species that exist or have existed. What kind of box or slot can we place humans in to understand them? We are the *only* animal to have such behaviors.

Before we begin to make up inexcusable reasons for who we are and how we behave, every creature documented on earth is classified by a set of scientific criteria. Science must follow strict rules for grouping, naming, and classifying living creatures based upon physical, mental, and behavioral characteristics. We are complex creatures, but not simply because of our physical makeup. The only thing that is consistent with humans is the fact that we will act and react to the world at different levels, at different times, and for different reasons. This is why we strive for knowledge.

One fact remains similar among us in that we do have the ability to consciously and routinely act against our innate instincts. Other animals, on the other hand, react according to instincts within their genetic nature. And that segregates the human animal from the rest. So how many characteristics must we have to determine what it means to be human? Perhaps acting against our innate behavior has made us human. This would certainly qualify us to drive and influence our own evolution. So our new beginning becomes very complicated.

One characteristic that is connected to human nature is our strong social network. This means more than simply living together as a group for mutual measures. Many species live in herds, packs, pods, flocks, prides, or colonies. Humans live in groups, including immediate family and desired social communities. They also live in geographic states, which many proudly segregate themselves from others with distinct, clear borders. And that's OK. People need to be content with that which they call home.

We continue to interact socially. People need to respect others' opinions, feelings, and ideas so that we can be harmoniously content. In doing so, we interact with one another to share and enjoy our lives together. Perhaps this is one of many aspects of why we are only human.

Other animals, however, congregate together for different reasons. They may inadvertently assist one another in an attempt at survival. Or do they? The social aspects of dogs and their wild counterparts, wolves, as well as whales, lions, apes, and other species, show a need to associate and interact with their own kind. There is no animal other than the domesticated dog whose complete desire is to be with, protect, and receive the loving affection of his master. This is a rarity based upon its evolution.

The compassionate human is more than willing to offer whatever it takes to care for this lifelong companion. The ultimate bond is between a boy and his dog. There is no questioning the affection in a dog's eyes when they look at you. But humans are very receptive to bonding with many different kinds of animals. It is as if there were an innate need to care for and protect living organisms other than our own kind. It is socialization at a different level for different end results. A new beginning is moving rapidly. We are human, so therefore we are social animals.

With physical, psychological, mental, and emotional changes occurring in human evolution, a well defined image of our species emerges within our recent history over the last fifty thousand years. Somewhere between this time frame and one hundred thousand years ago, there existed several different species of humans on our planet. These three to four species were *Homo erectus*, *H. heidelbergensis*, *Homo sapiens*, and Neanderthals, a close cousin to modern humans. Each had different skills and qualities to contribute to their own kind. Competition among these species trimmed humans into only one remaining species currently existing today, *Homo sapiens*.

However, during this time period, each contributed, often through their own sacrifice, various qualities and behavioral characteristics to the human race. We learn through example, we learn through imitation, and we learn through sharing information with each other. Those remaining hominid species competing in a vicious world gave each other, knowingly or unknowingly, skills that modified or changed their social networking or social behavior as a group. Nature is designed so that not everyone survives.

It's claimed that the Neanderthals were the first group of early humans to bury their dead. This ritual is estimated to have occurred some thirty five to fifty thousand plus years ago. Remains of Neanderthal burial sites indicate that the deceased were meticulously positioned in carved out gravesites, occasionally with Neanderthal artifacts that served as honorable tokens of respect for the dead. These so called primitive humans had demonstrated that they worshiped their life and respected their dead companions. A higher order of thinking and deeper ability for reasoning and feeling was leading us toward true humankind. The animal was becoming less of an animal and more of a human.

Human nature was changing as fast as the physiological variations among our ancestors. Over a short period of time, the world was beginning to appear different. Our expressions and sensations changed as our natural behavior evolved in front of us. From one aspect, the world was getting smaller in terms of common tradition and behavior. Human ancestry was becoming more unified, and distinct behaviors were emerging. The new beginning had begun!

How did our early ancestors view the changing and expanding world in front of them? It is like a similar scenario in the verbal exchange between Captain Barbosa and Captain Jack Sparrow in **Pirates of the Caribbean: Worlds End**. "The world used to be a bigger place," claims Barbosa. "The world is still the same, there's just less in it," claims Captain Jack. In fact, from one perspective, it may appear that there is less in it, but that may be because there is more in the world today that is coveted among many. Where are the new adventures to be found? What are the new ways in which the world is changing for life today?

Much remains to be conquered. This would certainly include physical and emotional challenges. At least now there are common traits within our nature that have the potential to link us closer together. These include the numerous traditions that we simulate between cultures with personal modifications or different twists. We are social animals with customs and traditions. These behaviors remain among our own kind. Customs and traditions are what we depend on when all else appears to fade from our sight into a forever changing world. The world is still a *big* place, but there are many hands in it.

Although aggression and less humane acts toward one another have always been demonstrated between us, the focus for our ancestors stemmed around the need to survive. These are innate instincts that had little agenda other than the attempt to live another day. But collective goals and ideas drove our actions toward a common objective, survival of one's tribe and immediate family. We quickly began to realize that in order to continue onward, we needed one another. Personal satisfaction, greed, power, and manipulation over others had not yet surfaced within our nature.

But the beginning of seeing life as more than how it appears on a day to day basis grew as our minds began to wonder about the world they so desperately tried to fit into. A sense of purpose and increased respect for life emerged and grew within us. Our history was beginning to lay boundaries to the nature of what humans are, and they were becoming new.

History is a story about the past. It is a story based upon facts, observations, or experiences, generally by those who acquire perceived truthful information about the events of people or places in time. At least we hope that the information represents the truth to the best of one's ability. But I think some may have been written by the winners, not the losers. Each one of us is part of history, although many are not written, glorified, or publicly honored. History is layered with information of different importance or significance. Human history consists of years of collected presumed factual events and significant family ventures and noted tales. However, the important thing is accurately recalling, recording, and telling the truth of an event. The value of personal history is in the telling of the tale so that it can be passed on. And let's face it, there is nothing more *humorous* than family history.

Not surprisingly, we fondly remember or share the events of such history at specific periods during our lives. One such time is during the eve of the New Year, which is often celebrated to the full extent. People in general look forward to new beginnings and putting behind some events of their past. It is also a time to hold onto one last glorious event that occurred during that year. Time may move forward, but memories have this ability to allow us to relive the past and may last one's lifetime. Collectively, the good and sad moments make up our individual history. In the end, a life with countless memories reflects a full life, certainly not one wasted by fear of venturing into temporary unknowns. But the real value of your history is how many tears of laughter you can give to others. And that can initiate a new beginning for all of us.

Enter the notorious New Year's resolution. Each and every eve, a personal goal is openly professed by some to better oneself or toward others. I think liquor may have a lot to do with it. This is an example of pure optimism and hope, assuming we can remember those commitments in the morning. Then, of course, there is the denial that we ever uttered those foolish words of repentant and goal oriented intentions. Many believe they are too fat, too lazy, unmotivated, procrastinating, self indulging, undisciplined, or less serving to others than they should be. That may be true. Then, on the other side of that coin, it shows a desire to better oneself. Don't beat yourself up too much or expect that you desire more than you already have. You're only human.

I don't commit to the yearly resolution gig. Like Don Henley of the music group the Eagles sings, "Get Over It." I try to resolve my ways more frequently throughout the year. It's also less publicized. And I'm certainly not going to declare any goal that is way out of my reach or unrealistic. You don't want to let yourself down too hard. So chocolate is still a routine part of my diet. In fact, I've become so realistic that the five basic food groups may still include sugar, fat, grease, starch, and alcohol. I will throw in some of the healthy foods that include vegetables, except the green, yellow, red, orange, and white ones. Seriously though, lighten up on the expectations you place on yourself. There are plenty of demands that life has for you already.

With all of this being said, our ancestors had slowly become enlightened. This demonstrates a sense of belonging among human relationships, such as various social customs, including the ritual of burying the dead, drawings discovered in caves, personally crafted jewelry, and many more artifacts common among early ancestors. Artistic creativity was becoming very instrumental in our society. Although specific burial rituals have changed over time, they have become customized within many cultures to the current day.

We cannot run and hide from our history, nor should we. Nor can we deny certain historic events because we wish to avoid pain, humiliation, or defeat. On the contrary, we need to embrace cultural customs. We should respect different customs to learn from and overcome our biases. Social customs hold great traditions of hopes and dreams for all. There isn't just a new beginning from each of our past species, but new beginnings from generation to generation. Our ancestors started to see this. They learned by living as a tribe and became more human. This is evident that we are only human.

I am continually humbled with each failed attempt that I've undertaken to not recognize the importance of someone else's traditions. Perhaps it is only because I don't always understand the customs. I respect them. But not fully knowing their history makes me ignorant of their importance. It simply means that others and I need to know more and respect other cultures' customs.

However, optimism, stubbornness, or courage to try again is what keeps us going. I believe our past ancestors were the same way. Work harder at acceptance of other people's intentions, and we may just become more forgiving and not hold onto discriminating views. There are some world concepts to accept and others to ignore. I would hate to think I'm too damn dumb to just let things go. But there is a fine line when it is time to leave something alone. It is this and a million other qualities that define human character.

If we want to make a new beginning for ourselves, then do so. To understand ourselves even better, we can choose to view our differences or our similarities. The choice is yours. I recommend looking at our similarities. We can also choose to view the positives or the negative qualities people possess. Humans must reflect on successes and failures to understand true strengths or weaknesses. We all have them. And there are times when we must make difficult decisions. Choose wisely!

How we see ourselves and behave will define our character and create new stories for future human history. The trick is to know what to cut loose and what to never let go. The disheartening loss of the Neanderthals some twenty eight to thirty five thousand years ago is balanced in nature by opportunities for the survivors. It is always sad to lose any species, but that is life. It is a necessary part of evolution. Nature demonstrates that life will always find a way. It is harder to stop life than to start it.

And the evolutionary progression of humans has enabled our last and only species, *Homo sapiens*, to accelerate the growth potential and exploratory accomplishments of the most unique animal ever to populate the world. The emotional and mental changes that have occurred in our species have influenced our current human nature. We are what we are because of how we respond to the world through our emotions. Mental intellect or mere physical prowess does not drive human behavior or describe human nature. It is the expression of feelings from emotions coordinated by our brain that often best illustrates us. Perhaps it is our emotional responses that are an important factor in what it means to be human, which is a comforting thought. And this would be a new development.

It would be wrong to assume that our early ancestors were merely heartless savages with emotions that were unkind or less than gentle toward one another. Similar and different species always compete. However, there was a point in human history when our emotions began to drive our behavior other than under the direction of our natural instincts. A whole new beginning had emerged as feelings and emotions about beliefs and freedom became entwined among our early civilizations.

Our earliest ancestors searched for new beginnings in an unknown world out of their own curiosity. This innate quality strengthened their ability to survive. They found new beginnings out of necessity and desire for exploration. Our ancestors changed from being gatherers to scavengers and eventually becoming hunters and farmers. They were constantly implementing *do-overs* until they got it right. Courage and confidence started to become part of human nature. The weak didn't survive, and the very old faded away. These strategies almost feel as they do today. But these are the harsh realities humans face. As much as the emotional state of human nature may fear change, new beginnings are inevitable and necessary.

Chapter 2: Life Calls to Life

When I intended to write this book **We Are Only Human**, I needed to consider the complex content. I certainly don't claim to be an authority on life or human nature. I taught high school evolution with biological anthropology along with human anatomy and physiology to sophomores and senior students. My approach to this book had to be logical and progressive in its content. This progressiveness had to include early human species of development brought about by evolution. We can very well see differences among our kin, but more importantly, the similarities that occur today.

The first chapter may have appeared to be somewhat of a textbook-like format of information regarding our early ancestors. But, with much thought and hesitation, I was determined to write about myself and my family's experiences over the years. How could I try to explain what it is to be only human if I didn't share my family's life experiences? In doing so, it occurred to me that almost everyone I know has had experiences very similar to my own. Perhaps we are not so very different. I will admit that as I have progressed through this writing, there have been so many humorous and hysterical events rediscovered over the years in my lifetime. And I am so fully satisfied with that. Therefore, with forever gratitude, I am extremely thankful for all of you. The closer one looks at all of us, the more you may see that life always calls to life. And that differences aren't so different.

Humans are not solitary creatures. As often as we may say "I need to be alone," it's usually limited or under specific circumstances. Try understanding this concept with a mom feverishly juggling all the kids off to school, remembering who has soccer practice, which one has swimming lessons, what time is the piano lesson, and occasional dentist appointments while shopping at the grocery store for a family who consumes food at an exhausting rate. Moms seldom have their "me" time. They certainly deserve it! Dads too cannot be excluded from this. I admit my "me" time is finding a quiet place to write without being rude to others at home. When the family is home, we do things together. I would feel somewhat guilty if I monopolized time for only myself. I easily spend hours writing or working on music. So, if my wife and kids are fine, I will take that "me" time to write or work on new lyrics for a song. In fact, many times, a husband and wife on a rare private dinner date from the kids often wind up talking about the kids they've momentarily escaped from their hectic responsibilities. Moms, not dads, begin to feel an unwarranted guilt when they get those infrequent "alone times" from the normal day's events. There's nothing wrong with the idea that "I need to be alone" once in a while. I'd never be able to write or compose music otherwise. That is why midnight, when all are asleep, is my quality time to compose or write. Fortunately, people tend to be tight-knit with family, which is what makes our kind unique among many other high-level creatures of the world. The difficult part is often finding that balance that satisfies the family tribe as a group and making time for your own personal needs.

We live among our own kind because it is inbred within our makeup to do so. Animals of prey have a better chance of surviving living within a herd. There is safety in numbers. There is no particular advantage in being alone. Predators select the old, weak, or injured among the herd during the hunt. They are targeted for the kill. It is nature's way of selecting one for the good of the many within the herd, and the continuation of the predator groups. Aside from their innate attitude to live together, do they also congregate out of simply the desire to do so, such as us? I believe they very much do. However, unlike us, they don't require their "me" time. Look at wolf and lion prides in the wild. Their "*play*" among them is social bonding. It is one of many examples of life calling to life. People live out of the need and the pleasure of each other. There are benefits and advantages to this way of life, but our desires and wishes often override what nature installs in us. This is what makes human nature different from an animal's natural behavior. Animals have no personal desires or wish for themselves as humans do.

Around the globe, people strive to live among their own kind for the comfort of psychological and emotional security. I believe altruistic behavior to be part of our human genetic makeup. At the same time, each of us has a unique niche requiring a certain amount of space to call our own. It is a domain that we determine to be our personal territory. As much as we need each other, there is often time for solitude for contemplation and escape from the day's realities to self-recognize that we are still individuals and independent for free thought and actions. So, the next time someone says to you,

"I need to be alone," don't take offense unless it was more like, "Get the hell out of here for a while, will you?"

* * *

Growing up in a family of five individuals with my two brothers in a modest-sized house had its moments. Being the oldest by five years, I, of course, had my own bedroom. The challenge in our house was sharing only one bathroom. Three boys can easily stand around peeing in the toilet. This certainly was more difficult for my wife, who, when growing up, lived in a similar house but with four other siblings. The bathroom all of a sudden looked smaller than it really was. However, everyone adjusts with few complaints. There are much larger families with fewer accommodations in the home. The noteworthy problem I recall was whose turn it was to empty the trash, cut the grass, or who conveniently forgot to put a new roll of toilet paper in the holder. Everyone among the neighborhood families had similar issues in their houses. There is a uniquely common behavior, without being taught, that transcends down the generations: should we leave the toilet seat up, or put it down? I'm still not sure.

No matter which background one may originate from or what their circumstances are, the frequent demands, pace, and expectations we have require us to call out to others of our kind to fulfill or accomplish our desires. Life has an extraordinary way of accomplishing this. We pick up that non-organic, electronic-dependent device seemingly attached to our hips called the cell phone. Saving 0.75 seconds, we speed-dial our wives and noted saviors to rectify the crisis and avert catastrophe. It avoids near panic mode to remind our spouse to bring home ice cream from the grocery store as she continues to load up the cart with nothing but the best from her five basic food groups, not necessarily mine. It doesn't matter how trivial or significant the crisis is; most humans often react in the same way. And once again, life calls to life. It's just another prime example of how our species continues to contact and depend on one another.

On a more serious note, the greatest opportunity to witness life calling to life is your newborn child crying out for *your* touch, *your* words, and *your* comfort that only a parent can provide. Infants will cry for several reasons. One is simply to be recognized that they, too, are a part of this endless life cycle. They must exercise their little lungs as their brains begin the process of interpreting the world. They are astute by various sounds and the responses made around them. I relate this to birds singing in the clear, still morning air during spring and summer. As a biologist, I would be obligated to explain why birds sing: simple communication, mates, territory, and the like. But I also feel that they sing because they can. There's nothing wrong with doing things that you're good at and enjoy. And although I have never asked a bird, part of me believes that birds sing because they are drawn to the rhythmic sounds of their own music. All animals, including the nature of humans, have a song to sing as life calls to life.

Fortunately, life often calls at the right time, when we are at our greatest need, seeking the most help from others. People innately call to each other in times of crisis. Groups of individuals do not hesitate to assist those in desperate times. Throughout the year, during hurricane or tornado seasons, thousands are in dire need at multiple levels of assistance.

Help often comes from around the country. Even with disasters that occur around the world, people are willing to lend aid. Sometimes our minds are as open as our hearts. Life continues to call to life.

With so much conflict and pain going on almost everywhere, individuals find different and sometimes unusual ways of protecting themselves from the horror and agony that each of us must endure from random untimely occurrences in our lives. We seek out solitude and sanctuary. For me, the passing of time through the natural climatic seasons has an uplifting and often reflective quality that helps me connect with others. There is nothing in this world like the spirit of the Christmas season or summer with my children and grandkids during days at the beach or barbecues at home with the entire family. These routine events have renewing qualities. Therefore, looking forward to specific seasons or social gatherings brings delight and hope for anticipation in various forms of joy.

I am fortunate to experience all seasons living here in the Northeast United States. I don't mind the heat in summer or the cold in winter. I know that may sound a little strange to some, but being a part of this life, I want to feel and experience all aspects of the seasons. In the spring, the birds return in mass. Their sounds in the predawn morning wake me from sleep. The sun has yet to rise and they are singing. If we don't wake up singing, at least greet the day with a little optimism. How might we behave for the remainder of the day if we did so? As we drag ourselves through the morning hours, we call out to life in various ways. Some are morning people, vibrant and full of energy; others are a little more solemn or more like *"get the hell out of my way until I have my coffee."*

Whether we wish to or not, communication is built into our nature. That behavior certainly differs among us. I find it amazing that it isn't only with people but all living creatures who have daily habits and routines.

During a summer's eve, the sounds are filled with locusts screeching out into the stillness of a dark night. Fireflies illuminate hidden corners of the yard among the greenery as if magical powers enabled them to instill wonder in a child's eyes. And crickets, using their hind legs, create such a unique sound as it echoes through the valley. These sounds of summer remind us that we are not alone, even if we choose to sit out at night to gaze with calming wonder at the stars above. In doing so, connections can be made, in the fact that the variety of life has much in common. And that is a most comforting thought as the tremendous abundance of life calls out to life.

I can recall as a young child, living at home one summer afternoon when we had a torrential thunderstorm. As kids, we all enjoyed the excitement of lightning and loud thunder during a hot summer day. Even now, my wife and I will sit on the front step or go out back to watch the approaching storms. Of course, the dogs are also with us. Well, that one summer afternoon, so long ago, a storm came upon us suddenly. Everyone in the house ran around like the Three Stooges, closing windows and doors to keep the rain out. I went upstairs to look out the window at the rain and saw a large bumblebee trapped against the outside windowpane, getting pummeled by pelting rain.

Suddenly, the bee was washed down to the ground and into gathering puddles of rainwater. I knew that little bee was in trouble. Without hesitating, I ran outside with a paper cup to find him. The downpour of rain and gusting wind made it difficult to see. I desperately looked for the bumblebee in the saturated ground. I was drenched to the bone. Among the pouring rain, I saw a little movement. Finally, it was the bee in the wet mud. I picked him up with a paper cup and brought him to a dry area out of the storm. The bee was motionless and looked dead. I remember feeling really bad and surprised at the sorrow I felt that he was no longer alive. I stayed with him for several minutes out of the rain until I saw the bee move, its one wing. Then the leg, and finally, both wings slowly began to flutter. The storm passed as quickly as it came. I then saw a little more movement. The bumble slowly gained its strength and began to walk around, attempting to dry off its wings. Eventually, I saw him fly away into a clearing sky. It may have been a small thing, but nonetheless, an important life to me. Life calls to life in so many unique ways.

This happened over forty years ago. I remember it as if it were yesterday. Saving that bumblebee's life meant more than any harm that could have occurred to me out in that rain-driven lightning storm. For me, it was a natural thing to do. We all act without hesitation at times, offering ourselves for the sake of others. It is a human thing to do. Or are these things part of what makes us human? That bee spoke to me in a way I can never truly articulate. I can only hope that there will be more events that qualify me as being a noble human toward life. It's our responsibility. Some may laugh, others may agree, but I'm all about owning up to responsibility and purpose. Every day, I think to myself how I can justify my purpose in being here. A person's personal purpose varies among us all. When we look at others, some may not think those duties are relevant or that they're inconsequential. But who is to judge the value of another's worth? Humans may be cynical enough to measure others. But we're not all on the same scale. Most don't understand the value of their own life, let alone others. The human scale of value has no numerical worth. It becomes a balance of equal rather than a measurement of scale. And the accomplishments should be personal and not of a hubris nature. Fortunately, much of what people do is not done for themselves alone.

Even in death, life calls out to life with unspoken words. Honor those of the USS Arizona whose sacred tomb lies in the waters of Pearl Harbor. The tragic loss of life from the sinking of the Titanic reminds us of the hubris of people who once believed that our abilities were indestructible or invincible. The unimaginable accounts at the twin towers in NYC. The dead and dying in Africa, the Middle East, and Asia call out to each of us about how thoughts, opinions, and behaviors must change for the sake of humanity. This is far from being a silent world, even when voices are not spoken. Their calls can be heard in different ways. The questions might be, how can we put personal ambitions and prejudices aside and do what is right for our kin, who are so genetically alike, and that there are virtually no true differences between us? Life is still all around us.

Life can call to life without being spoken throughout the universe. I don't think we need to ask ourselves the question, "Is there life out in the universe?" I believe we need to ask, "How do we get to it?" Our universe is *primed* for life. Evolution has shown us that time and change enable some species to adapt and thrive, while others are replaced. Life moves toward perfection that the current environment will support. Therefore, as the evolution of all species continues onward, how will life call to other worlds? Is that a purpose for humans, and should our behaviors be more open to such concepts? The baby bird must someday leave its nest. How will we leave ours, and on what terms?

Science calls out to life daily as we search the heavens for the existence of living organisms and new wonders that may enlighten us to the meaning of everything. Life called to life five million years ago within our evolution and continues to call out today. That was the time period when primitive human-like ancestors emerged. We certainly didn't become human until much later. It may very well be from physical and behavioral terms or from a biblical nature when humans came to be. Perhaps it is a combination of both. This is a tale and discussion all its own.

For us to survive as a species, we must pool our resources and lend a helping hand so that our planet can survive. There must be leaders, followers, winners, and losers. However, we cannot be blind to life around us. The world, from a completely different point of view, is getting bigger, not smaller. It is no longer the survival of a small group of ancestors. It is our compassion for our species to pull together, working together, and helping one another when possible. If we are to endure the realities of this life on our planet, we must act unlike any other time in human history. It is life calling to life in a unique and profound way. Our own living earth calls to us to be saved from our negligent behavior. This is a calling to life unlike any other. It is a way in which many of us hear it today. Although some may hear this song, it is important that we all listen during this critical moment in time. We have one and only one earth, as we have one and only one human species.

Therefore, did our ancestors leave Africa over a million years ago because of the need for food, space, and various resources? Or did life call to life from a perspective of human *curiosity* and need to explore new lands? This may not have been only for survival, but for navigating into a future to become the most unique form of life on earth. When did humankind limit calling out but loudly proclaim, *I AM HERE!* Was this the beginning of a new human-based element that had never been seen before? Perhaps these human drives are built within us genetically so that we can survive and continue onward. Whatever the reason, it has become an open expression of our behavior to not only survive but to be recognized, as life continues to call to life.

Chapter 3: It's a Nor'easter

Living in the northeast United States for all my life has provided some skill in dealing with an occasional severe weather condition called a Nor'easter. Storms from the northeast can punish the coastline and interior from below New Jersey up to Maine. Three days of torrential rains and winds during the summer can not only upset the vacation plans along the shore for young ones, but it can also create devastating destruction to property and homes. Today, we have structures and action plans to help cope with such destructive forces. Our ancestors years ago had little hope of ensuring their survival when struck by hurricanes, tornadoes, or other horrific storms. The collective recovery actions by people today come in a variety of outpouring aid to those in need. Thousands of years ago, fear, confusion, and instinct from such events forced our ancestors to regroup and carry on. Needless to say, the horror of a Nor'easter can have great psychological effects under certain conditions.

As a kid, I practically grew up at the shore in New Jersey, visiting my grandparents' summer home. Every year, work on the house needed to be done before the summer season began. It was an entire family event to prepare the house for the summer season. During early spring, the outside of the house always needed painting, cleaning, and de-winterizing before our summer excursions. Afterward, we were free to enjoy our time along the shore. Even at the age of three, I wanted to help mom, dad, Pop, and Grandma fix up the summer house. To satisfy my wishes, Pop—my grandfather—would get me an empty coffee can filled with water and a paintbrush for me to "paint" the back porch steps. It went on like paint, so I never knew the difference. It wasn't until the age of five that I was taught the proper way to handle a brush and can of paint. By then, we all worked together, and by late afternoons during warm spring weekends, we'd walk to the beach. My family instilled in me and my two younger brothers that there was time for work and time for play. It was a lesson that I am grateful to have learned.

There had always been cautious trepidation when potential Nor'easters were threatening or approaching. We took the destructive power of such storms in stride. There was no other way. The integrity of human beings is to rebuild, repair, and move onward. We would never just give up. Being at the mercy of nature was nothing not to fear, but was respected and gave way when possible. The resilience of people seems greatest when situations are at their most challenging. Human nature over the years has hardened us for such battles with nature. The internal strength of people is incredibly strong and often unwavering with great resolve. It is a characteristic of humans which is not admired nearly enough.

Winter Nor'easters in the northeast United States provide for significant snowfall under the right conditions. It comes with a smile on the faces of kids for a day off from school along with limitless sledding and snowball fights. Different parts of the world have their climate challenges with even more severe obstacles to overcome. But because humans have grown from a population of approximately one million people ten thousand years ago to currently eight billion, we now inhabit all types of terrain out of necessity. Our evolutionary process of change has enabled us to adapt and utilize our cognitive skills to shape the spaces we live. Without such abilities, we would have dwindled to pockets of humanity over much less area. However, no matter where one lives on earth, Nor'easters and other severe storms are continuous challenges.

Subconsciously, I think humans wish to go where the wind might take them. Very much like a kite flying freely, even with the inherent risk of a devastating crash that could end the journey. People love their freedom and the outward venture of pushing safety boundaries. As much as human nature guides us toward desired paths, the unknown and risks associated with danger draw us into those rollercoasters of emotions between safety and blind adventure. Early exploration of the seas fueled our need to accomplish peril, as well as our curiosity. Human behavior is inundated with individuals who explore the depths of the ocean, climb the highest peaks, or seek out the fastest means to travel. We also dream about venturing into the universe. Those types of thrills found in life satisfy and, at times, calm the soul. A trip to the amusement park once a year to face the challenge of the newest, fastest, heart-pounding rollercoaster takes us to the brink of potential disaster and fear and then elates our senses and emotions with courageous accomplishment. For some people, the amusement park does the trick. For others, a more extreme challenge is required to fulfill their pleasure center. In any case, the nature of people is to continually move in a specific direction that often changes courses based upon needs, desires, and unexpected events. All people want freedom and to choose for themselves that they can almost taste it. It's human nature.

I have always admired seafaring individuals, particularly those of old. They were always aware of potential storms at sea but traveled the oceans anyway. Great courage and internal resolve were needed during those times. There was no GPS instrumentation to guide their way or chance for rescue by others. Early explorers showed true strength to face such risks. Seafarers facing a Nor'easter, a typhoon, a hurricane, or unpredictable conditions of the world's oceans embedded admiration and gratitude for their efforts by many. We can list several motivating factors that may drive people today to risk life, but back in the day, it was raw strength and courage. All they needed was the wind in their sails, a compass, stars to guide their way, and a crude map to familiar and unknown territories. One cannot help but envy such adventurous determination. Over land or sea, humankind's exploratory travel has made its mark on our continual behavior, which makes us who we are.

So which direction is humanity traveling? It is human nature to explore. We seem to settle in an area and then eventually gaze over the horizon. Give thanks to our instinctual curiosity and unshakable need to survive. But is there a Nor'easter out there just waiting for us? It is more likely probable. How would we handle it, and what course would we take to steer clear or collect the fallen pieces? We must look at our history so that we may learn from our mistakes. Human behavior begins with instinct and expands toward learned experiences. Sadly, many are inadvertently sacrificed to destructive forces of nature. But we, as a species, go on. However, this is the way of life. Other animals fall into these circumstances. Fortunately, humans have an innate motivation to help those in drastic need. We simply must expand that effort beyond assisting humans. We need to recognize our altruistic instinct to help save all life. For the road we travel impacts all other life forms on our planet. To predict our future's direction, we should look at where we came from.

We go back in time to the predawn hours on the African Savannah a million years ago. An intelligent master predator emerges. A small, organized band of early humans gathered and undertook a trek northward into unknown territory. Their mission was to seek out new territories for food, water, and the discovery of what lies beyond their realm of visual sight. Curiosity and basic needs drove their risk to expand their known world. The challenges were great, the rewards uncountable, and dangers were vast beyond their comprehension. Yet, our ancestors left Africa and migrated into Europe, east into Asia, eventually across a land bridge into North America, and finally into South America. They were as brave as the sailors ready to battle whatever storm came upon them.

It is even believed that some of our ancestors much later took the risk of traveling from Europe further north and then west across ice sheets and open waters in primitive sailing crafts across the Atlantic Ocean to the eastern shores of North America. Many died during the journey, and yet, beyond their fears and uncertainty, they continued onward. Any direction one travels will always lead somewhere. Riding that trip through time will immerse us in both calm and rough seas.

The peaks and valleys of pleasure and sorrow come with the territory. Nonetheless, we all must take the journey as our ancestors have. However, becoming aware of our path's direction can help us deflect disaster or hit that wall of water head-on. Another Nor'easter will come.

It requires courage and instinct for any species to travel outside their current boundaries. Instincts to meet basic needs for survival are powerful forces built into an organism's genetic makeup. But there is no known gene for curiosity. Human behavior is full of curiosity. Our needs force us to move out and explore. We travel through the lands, the sea, and now into space. Perhaps it is destiny or fate that drives us to seek out new frontiers. It is the need to survive that drives the beasts of the world to move from territory to territory. The migration patterns and behavior modifications of many animals are innate and determined by genes. They are influenced and triggered by changes in weather and established as an adaptation that assists in their survival. The need to emigrate or lie dormant may be equally as strong. This is true for many different species.

It can be explained why some animals migrate and others immigrate to new areas. The migratory patterns of birds, butterflies, schools of fish, caribou, water buffalo, and hundreds of others ensure unity and balance in life. Nature has done a great job making available the different adaptations for all species. Survival of the fittest may be able to justify acquired adaptations in which species continue and many become extinct. But I find it fascinating why nature provides all the tools for life's survival. And yet, it takes away those necessary elements for others. It's as if life must go on no matter what environmental circumstance arises. Why doesn't life fail altogether? It doesn't, and it won't. By all rights, we should have become extinct long ago. We were on the edge many times in our distant past. My curiosity is not how something happens, but why it does. And species movement and settlement play a large role in overall survival. Even today, our yearly vacations take us away to a different location. If we can't physically travel, then the next best thing is to do our traveling in our minds.

It is the call of freedom.

* * *

Humans today travel for the "fun" of being on the road and reaching desired vacation destinations. That extended "family time" on the road, only to reach that coveted family vacation spot, is worth any bump in the road. We need to get to Wally World! Some may argue about the "fun" in a long road excursion. The carload of kids each asking, *"Are we there yet?"* Priceless! If you're the driver of the vehicle, it may make you have doubts about this "family time." Then comes the question, is it worth the multitude of circumstances that happen on one's travels? The family horror stories are endless, but at the very least, highly entertaining to hear, especially if you were not part of the experience. Once the stress resides and you accept the circumstances of the master plans, you quickly realize it was worth every second.

I can still remember the trip to Niagara Falls, approximately eight to nine hours from home, many years ago. Our daughter, before our son was born, was two years old and directly in the middle of "potty" training. For parents, potty training itself is a storm of its own. I'd have to estimate six hours of pee-pee breaks in addition to the eight-hour trip. Finally, on the way home from Niagara Falls, we wised up and had her tinkle in her potty chair in the back seat. We didn't even slow down below sixty miles per hour to toss that pee right out the passenger window. We thought, enough with the rest stops. Let's just get home so we can have a break from our vacation! It really wasn't that bad, and we continued to laugh about it many years later. They are memories we'd never choose to lose.

Several years later, my son came along on our car trips out to Dayton, Ohio, to visit family. Two young kids in the back seat. This became a Nor'easter of a whole new breed. Travel became much more challenging. My daughter, at this time, acquired motion sickness while riding in the car. Always writing or coloring in the back seat at the age of six or seven did not help alleviate her car sickness.

Many years ago, on our monumental car trip from New Jersey to Ohio to visit my sister-in-law and family, I knew this would be a memorable adventure. Several hours into our excursion, I asked my wife, "Where are we? Did you see that last road sign back there?" She replied, "No, I thought you knew where you're going." "Mommy, I don't feel so good," my daughter shouts out. "I gotta go pee," came the other voice from the back seat. "Let's stop the car for a break," says my wife. "No, we gotta make time," I'd say. "I think I'm going to be sick," still coloring, of course. "I really gotta pee now," says my son. I'd say, "There's no place to pull over. We're on the Pennsylvania Turnpike. The rest stop sign said another thirty-four miles." My son says to everyone, "I can't wait that long," says my son. "Blah" goes my daughter in the back seat as she vomits over everything within a perimeter of her projectile range. My wife says with such calmness and restraint, "I told you we should have pulled over back there." Eighteen hours later, we reach our nine-and-a-half-hour destination site. My brother-in-law asks when we arrived, "How was your trip?" Really! I don't think you want to know. But we did it twice a year without hesitation. I often thought that a seafaring journey across the North Atlantic Ocean during a hurricane five hundred years ago inside a small, cramped sailing vessel would have been a pleasure cruise compared to some of our vacation trips by car.

That pungent odor that lingered in the car from our family vacation excursions over the years was memorable. Between kids in diapers, to puke in the back seat during our intended destinations, it left lasting memories. We never hesitated on our family journeys, not concerned ourselves with possible outcomes along the way. We were a tight-knit family. One for all, and all for one. The "Three Musketeers" would have been proud! The journey is never more important than the numerous memories we'd acquire. Life may have felt a little cumbersome at the time, but executing the planned event was always worth it. Today, we reminisce about where we've been and what we've done together as a family. This is something no one can ever take away. Those little bumps in the road, whether they come from unforeseen events during a vacation or throughout life in general, feel like mountains when they happen but are inconsequential when you look at why we do what we do. As time goes on, they make for great stories sitting around the kitchen table. Family history is the best history of all.

*   *   *

It appears we learn to face our Nor'easters and disasters as we're forced to encounter hardships and overwhelming challenges. Our children come into this world helpless and dependent on us. Sadly, some of the elderly leave this world in the same manner. I wonder if it is meant by design for humans to establish the strongest bonds possible among us. Perhaps it is this driving force that gives us our resolve. There is no other feeling than the love of a parent for their child. These bonds among children and parents are eternal. It's a love so strong that it must last forever if nothing else ever does. But when did our distant ancestors begin to feel these types of emotional bonds among family?

We can suspect how various aspects were in the lifestyles of *Homo ergaster* and Neanderthals thousands of years ago, based upon artifacts discovered at communal sites. But true love for others or our children is hard to determine when that moment in time began for humans. Our emotional and outward reactions have evolved as equally as our bodies have over the years. And we continue to evolve today intellectually, emotionally, and how we view a higher spiritual power and the role it has in our purpose and direction. It almost appears that human survival continues only from battling and enduring fierce storms along the way. The development of human nature has many pages in its book. We will all face Nor'easters in our lifetime.

Chapter 4: It's The First Day of School All Over Again

They say history has a way of repeating itself. From certain points of view, that is true. The exact events differ, but the battles, the challenges, the successes, and failures seem to cycle in an unnatural loop. Is it fate and destiny or wisdom from our experiences that are needed to be understood or accepted before we characteristically move on? What is the trick to breaking the history cycle? Will we ever learn?

An optimist would suggest that humans are progressing forward in exponential bounds. Much of this is true. Don't listen to pessimistic individuals; they're not worth your time. Some of the basic innate behaviors and philosophies have remained the same for thousands of years. True human civilization behaviors and attitudes, ten to twelve thousand years ago, cannot provide a detailed accountability of human nature as compared to changes in civilizations over the last two thousand years. So, when, how, and more importantly, why has human behavior changed to include an endless list of what many would call misunderstood qualities about us? Much of this may be to deal with going back and repeating before moving forward to understand ourselves.

The first day of school can be the most joyful to some, and for others, a terrifying experience. For me, it was the latter. There are many uncertainties that are thrown at a young one going off to school. Not that these are given in any specific order of dread, but include leaving home, being away from a loving protective parent, meeting unfamiliar faces at a strange building, wondering when the superiors were ever going to let you go home. It also challenged your undeveloped skills in front of twenty-five other petrified little faces who were feeling the same as you. There is a laundry list of even more horrific aspects about the first day of school, which I have successfully suppressed over the years. School can be very traumatic. Why couldn't I have just lived in a primitive tribe and been given a knife to go out and kill a bear to become a man? It would have been so much easier. But no, we had the challenges of that scary building every young child knew as "school."

Thousands of years ago, human emotional development was at its beginning stages on the evolutionary timetable. Life then was hard, *very hard.* And that's a good thing. Because if we had to worry about such unwarranted stresses like we do today back then, we certainly wouldn't have had the strength to survive those important battles. So, was it by design, a little luck, or the mystery of nature's scheme that allowed us to develop how and when we needed to push for the courage to attempt something different? It wasn't a question for our ancestors. Their everyday routine was just to find a way to survive. Why couldn't I know that as a kid?

My wife always made a big deal out of the first day of school with our kids. She was generally excited and happy about their journey. So, she put on a brave face for them. Kids don't want to see mom cry at the door as they leave home, nor would they want to see dad hiding a tear. There always seems to be a fine line between sadness and joy. Yes, the kids need to grow up and venture out on their own. My wife and I never relished the thought of the day when our kids would leave for good. But even after college, they returned to visit often, thankfully, and life picked up right where we left it. We all live very close to one another and share our lives with growing families. And yet, the cycle returns as we both have heavy hearts when we see our grandsons begin their journeys off to grade school, as we still want to take away the worries of the world for them. Perhaps there's a vein of selfishness to hold onto people we care about. But isn't that part of being human? That type of selfishness, however, is the kind that expresses love, concern, compassion, and empathy.

The kids wore their new back-to-school clothes of their choice on the first day. Everyone showered, dressed, combed their hair, ate a good breakfast, and lined up near the door for a picture. Yes, we have pictures of every first day of school until they begin to no longer smile and see the humor in it. At least the short pants, suspenders, and bow tie outfit for my son was only reserved for Easter when he was four or five years old. We valued routines in the family when the kids were young. Our family ate together at the dinner table, read bedtime stories to the kids, no matter how tired my wife or I happened to be. Scheduled vacations included everyone in the family, and we knew what the plan was for each holiday event. These traditions were established within us from one generation to the next. It became part of our culture. Traditions play a part in the character of being human all around the world. It just may be the traditions that define who we are. Therefore, it was unquestioned that you were sure to get your picture taken during the first day of school.

Every year as a child, we have documentation of the "first day of school" for each consecutive grade level. However, this is not like the very first day of school when a child leaves home and parents to enter that large, mysterious building, which sucks up kids in the morning and spits them out in the afternoon. Nerves ran high throughout each grade level during opening day. Many teenagers hide anxiety by demonstrating their "cool" attitude among their peers and educators. But deep down, no child wishes for failure or wants the unexpected social pressure from teachers and other students during that first day of school. As a previous high school teacher, I have discovered it interesting and enlightening in how students behave in their school environment. When we dig down to the core of human nature, we are all so much alike in many ways. We're like a bag of gummy bears. There may be different colors in the bag, but they're still gummy bears.

Advancing into the working world after school also drags a piece of our past with us. Nothing is ever left behind. Am I the same gummy bear as before? This includes those unwanted past feelings and memories now stored in an area of the brain that prevents them from surfacing and dominating our current reality. Maturity or not, basic human emotions drive our behaviors. Sometimes I think apprehension toward new ventures is a good thing. It reminds us of what is acceptable or safe, which may be highly risky and dangerous. If this is innate, I often think of how nice it would be to genetically tweak that gene to allow us to feel more comfortable about some of the actions we must undertake.

So, as previously stated, we are only human. As if this were meant to be some form of excuse, or that which we truly are. But what does it mean to be human? There is no other time in recent history when we witness repeated events of speeches, promises, and actions, almost verbatim, made by our elected and want-to-be-elected officials. Especially during the time of political election periods. Talk about the first day of school all over again. There is nothing more insulting than to hear about one's commitment for changes that will be implemented for the masses when "selected," or should I say elected. Every four years, we hear the same verbal commitments about justice for the masses and positive actions that will be made for all by those who elect them. Pay me to write their campaign speeches. I know them by heart. Nonetheless, some people are easily fooled and want to hear what they want to hear. It's the first day of school all over again!

After so many years of continually trying to understand the various acts of "power" within people, I recognize that what is said, and the actions performed, are often for a select group. It is supposedly intended for the masses, but that innate nature to be the alpha and dominate the pack is frequently for personal gains of a control ideology. The human nature of politics belongs in a category all to itself. Sometimes the animal world, with alpha males and alpha females, runs more smoothly and consistently than it does compared to humans. Nature sets the rules, and genetic boundaries execute the outcomes.

Humans may disagree on every issue under the sun. However, civil debate, respectful opinion, and the ability to openly choose or vote uninfluenced by others should be a given right for any union of people on earth. This, of course, does not occur globally. A fine line in the sand is often drawn when we stop discussions. When honest and appropriate actions or sanctions are implemented, it is for the good of the whole. As often stated, *"The good of the many must outweigh the good of the one."* This concept may be one of the most difficult strategies for human behavior to play out. Differences in opinions and ideas about what is "right" and what is "wrong" have been disputed for thousands of years. This question of right or wrong may never be answered. Why don't we all take a picture of ourselves like the first day of school and look at it once in a while to remind us of where we came from and where we wish to go?

The question remains: how can we avoid ... "the first day of school all over again"? How can human nature adjust, or more appropriately, evolve to subdue the concepts of power and greed installed in each of us? History is full of documented events that include human conflict imposed by rule, domination, submission, and the desecration of human integrity and freedom. Perhaps humans should begin to view life on this planet as a whole and not simply as people who coexist. They would see that thousands of other living species that live here have more in common with humans than the average individual realizes. Maybe then a higher respect will emerge that could literally have universal changes beyond our current comprehension.

I find it interesting how some people strive to take great risks or gamble with their lives as adults. It is much more than simply confidence that determines their actions. For some, high expectations, ego, or a hubris attitude becomes part of their characteristic. In others, it is the need to explore and find contentment with the challenge, regardless of final outcomes.

I, at times, am somewhat envious of those who are more adventurous. Safe may have its comforts, but when, and to what degree, are we truly safe from harm? There are different ways that people evaluate security. Most consider the physical or financial risks. But the true harm is when we stop speaking to one another. Losing respect for others is catastrophic. Then you are never safe. As we grow older, our family home or personal belongings are also in the forefront of our desired protection. When we're older, one may think they've paid their dues, thinking they are smarter than they were in the past and therefore must be secure. Then, life comes along to remind us of what we don't know, resulting in more questions. When will I be safe again?

The question through life is: when do we take chances on things we may wish to choose to accomplish or try? In the back of our minds, we ask, Is this the "first day of school all over again?" How many times do I want to try this? Human nature tends to revert to old ways or customs when it comes to new or possibly risky ventures. Our intentions may be to go for it, but old habits die hard, and the anxiety comes back, just like our first day of school, to put the brakes on our actions. Just give me another day to think about it, and another, and one more.

With all of these factors being stated, how many qualities make us human? It appears that some form of moral code and actions that people undertake define their life. Each one of us is different and unique, but at the same time, similar. We may wish to assimilate the strengths of others into our personal lives but should not expect specific outcomes. It's only human nature to emulate others who we admire. Deep down, everyone wants to be better than they are. We learn from others, we help other people, we give to other people, but we aspire to be who we are to become an independent individual.

*　　　　*　　　　*

As uniquely different as every day will be, there are certain events that tend to be repeated no matter what. My grandparents frequently took me down to their summer shore house for days when I was a kid. To get to their home and boathouse in Wildwood, N.J., we drove on the Delsea Drive, the only direct road to our destination. My grandfather, Pop, would trailer his boat down to the boathouse in North Wildwood in the early spring. This single-lane road was heavily traveled. The ride was nightmarish, with very slow to nearly stopped cars due to heavy traffic. No one traveling to and from the shore enjoyed the ride. However, it was the most direct route to everyone's destination. The one-hour normal drive in the off-season took nearly two hours during the spring and summer months. Patience was very limited by all drivers, no matter which direction they were traveling. As young kids, we were only in a hurry to get to the shore, while for the adults, driving was agonizing. Even while I was young, I noticed how stressful it was to pack all the clothes, food, and toys that one would need for our excursion. It wasn't like we were going out of state or traveling across the country, but it felt that way. We were kids and thought and felt differently than the adults who had to endure the arduous trip.

Passing slower-moving cars on the way down was common practice for Pop. He would pass others while pulling his boat with the car. Pop drove like he was at the Indianapolis 500 Speedway. I believe he thought the brakes on the car would prevent any potential catastrophic event. I can still hear his voice as a little kid, *"I think I see an opening, Myrte, hold on,"* and off we'd go, passing cars with our boat trailing behind. It was a white-knuckled ride for me and Grandmom all the time. Those images of us rocketing in the passing lane are permanently etched in my mind. It isn't every day you see a car pass you while pulling a boat. This was a yearly event of getting the boat down to the shore. Some things just never change. These were memorable times which we still laugh about today.

The summer home, located in the center of Wildwood, where my two brothers and I stayed when we were young, is full of memories of absolute fun and adventure. All of us are now water creatures that have grown up at the shore, boating, surfing, and lounging on a sun-soaked sandy beach. In those early years, Wildwood was a family town with more to do than we realized at that time. There was also a lengthy "board" walk that had large piers extending toward the ocean with every possible amusement ride one could imagine. This was another huge attraction that drew people to visit Wildwood as far away as Canada. The summers at our shore were always packed with people. Fortunately for us, we did not need a hotel since we owned our own house in the center of Wildwood and boathouse with a pier and dock in North Wildwood. We were truly gifted to have such fun and adventure along with irreplaceable family memories. Year after year, going to the shore never felt like a routine chore. We longed for the time when school would let out and summer pleasures would begin. To this day, I am saddened when summer comes to a fondly remembered end. It is just as disappointing when it was necessary for us to sell the beach house and boathouse. It had been in the family for many, many years. These were great times as a child at our beach house. As much fun as this was, my family also loves the snow of winter but dreams of hot summer days to spend much of our time in the water.

While these repeat events occurred with precise regularity, they did make for the best of times. Our second property, so many years ago in Wildwood, had a boathouse and dock at the intercoastal backwaters of the island. Getting Pop's boat off the trailer, down the ramp, and into the water during early spring was at least a hysterical event. This routine would occur each and every year. People from neighboring houses would come out to watch the comic routine of us getting the boat into the water and around to the dock at our boathouse. People would come out during this yearly historic event with lawn chairs, sandwiches, and occasionally cameras to sit and watch us. My dad would be standing in the water and Pop would attempt, many times, to back the trailer down the boat ramp. To this day, I don't know how my dad didn't get run over. You would hear people yelling, "Turn the wheel to the right, to the right." "No, to the left, the left." I only wish that I had pictures of this to show my kids today. About an exhausting hour later, we'd tie up the boat at our dock, hearing the loud, hilarious cheers and whistles of the neighbors. A few beers later, Pop, Grandmom, Dad, and Mom were all laughing about it. The funniest thing of all is that it happened every year as if pre-planned.

Often, that "first day of school all over again" would leave precious lasting images in our minds. We ponder those memories fondly and revisit the events on a yearly timetable. Our thoughts and memories occasionally function as scheduled seasonal changes. Nature has a way of a distinct routine progression of spring, summer, fall, and winter, each with an impact on living creatures year after year. The same can be true for human actions. There is an expectation of what each season will look like and how it will make us feel. Recalling lasting memories rejuvenates the spirit and gives us the courage to begin again. Our three golden retrievers also enjoyed warm spring days lounging in the backyard, hot summer days to swim in the pool, and again to lay in the cool autumn daylight without the excessive heat. Animals may not be anticipating the coming of the seasons, but they do enjoy the changes that nature provides. Perhaps somewhere in the hearts of my dogs, they long for these changes which they always enjoyed.

* * *

There is such a fine line between all emotions ranging from utopia and bliss to tragedy and despair. Humans are complex creatures. This is why we are only human. At any given moment during the day, we are naïve with comfort and security in believing that we are not alone. However, in an instant, it can all be taken away without reason or an understandable explanation. Just like the days of past summers have changed, the same is true for life changes.

The striking blow of the sudden death of a loved one always leaves a lasting scar, no matter how one heals or copes. These battle wounds are evident even when we know that death for a loved one is near. Justification of such emotional heartache can never be rationalized by words. Hopefully, this can be accepted as a requirement to have truly lived a life. No matter what the circumstance for personal loss, there always remains the "why" questions as to the timing of the event, as if we actually had anything to do with the inevitable action of losing someone of the heart. Time may slowly heal loss, but that amount of time may seem to last forever.

As most of us attempt to navigate through obstacles that are confounding, we sometimes discover that we've attempted to resolve these issues before. Although occasionally successful, there are times when answers appear to elude us. We are then forced to revisit previous situations. How often have many felt that an experience has repeated more than once? Is it coincidental, or are our minds bending the perceived issue to related events that we've already gone through? Either way, our behavioral approach is often the same, as if we didn't learn the first time around.

Most seem to dread that *"first day of school all over again"* feeling. The anticipation of events that may have come as before. However, instead of them being troubles to solve, perhaps our minds are showing us similar obstacles to enable us to develop new pathways or solutions to those challenges. If we accurately solve them, that *first day of school again* feeling may not return. But don't be surprised if we move forward and then stumble backward from time to time. We are only human.

However, it is uncanny that certain events in human history repeat as if the book had been rewritten, this time with different names and characters to act out the same parts.

Nothing could be truer than that during political election years. How many times do we need to hear the same rhetoric with slightly new twists on past concepts, as if there were anything new about it at all? I'm more of a realist than anything else, but for the sake of being optimistic, I'll go out on a limb and state that most people who run for political office are moral individuals. It is practically a given that they wish to do good things for the people they represent. Human nature has a built-in quality of altruism, especially when we are at our most troubling times. On the other hand, human nature also exudes hubris, domination, and the need to exercise perceived power. If you wish to witness true power, try looking elsewhere rather than politicians.

To live in a civil world, structure, respect, order, guidance, and leadership are among just a few concepts needed within a governing system. Do not silence the people. Opinions, debate, compromise, and understanding of one another are a must. Charlie Kirk knew this! He believed in speaking, talking, and actually listening to one another. The natural order of humans is so complex when considering the human aspect of ambition and motives. Specific objectives and often sightless vision are difficult for us to respectfully accept. This may be only one thing that Charlie was attempting for us to understand. *Words* matter! *Actions* are equally significant! We are a species of social and closely interacting animals. It is inbred in us to be together. The real problem may lie with the expression and uncontrollable human emotional qualities that are within our genetic makeup. These delicate and valuable emotions we have occasionally cloud our sight of sound moral judgment. Nonetheless, without them, how could we be human?

Humans make mistakes, and humans make grave, devastating errors against life, but these appear to be underlying aspects of what humans currently are. In the political arena, men and women will passionately battle one another for what they believe in or how they want their supporters to view what they are supposedly accomplishing for them. In America, this "fight" for one's constituents is the expression of democracy as it had been intended. This is the greatest country in the world. And for a good reason! The majority is to rule, and all will live with the decisions mandated, or *until* a new decree is determined. Lately, more and more tainted objectives intended for the good of the people have been quietly introduced. Some people can be easily fooled. Always keep in mind that power only seeks more power.

We occasionally see promises and their outcomes geared to satisfy a select minority or influence members of government almost covertly. Often, these agendas are not in the best interest of the majority. Anytime there is the need to spend billions of dollars to buy a seat in American politics, it is ethically uncalled for. "It's the first day of school all over again." Power has taken hold of many aspects of human behaviors, orchestrating game plans not intended for the common people who starve for *real* leadership. Today, people are so desperate to make their lives better that they'll listen to the loudest voice constantly professing that which they desire to hear. When one is told something long enough, sooner or later, they begin to believe it. But I suppose that's human nature.

Human nature and certainly human behaviors have changed even in a single lifetime. And that's OK. As a biologist and observer of this species, I would expect humans to change behaviorally and hopefully with higher moral and ethical standards. I may be asking too much. These behavioral changes should be passed along to new generations. This is critical where family values play a significant role. Without a core family-based structure, most young ones fall to outside influences that do not align with respectful norms. Faith-based values are also important and self-satisfying. Confidence and inner strength are then bolstered. It is the only way in which we are to survive as humans. Unfortunately, there are many qualities, or lack thereof, that need to be addressed for all eight billion people to coexist. I don't believe things should remain as they used to but should progress toward a "fairness to all." There is a very fine line between "equally to all" and "fair to all." I often see people misunderstand the phrase *"equal"* and *"fair."* If you want to know how to evaluate someone equally, ask any teacher how this is done. Assignment grades cannot be given equally across the board in the classroom. It is absolutely unfair to the students. They must be graded fairly because of the multiple avenues of learning individuals have and in the evaluation strategies a teacher utilizes to assess that knowledge. If you think the terms fair and equal are directly related to the phrase *"freedom,"* then think again. That phrase is different altogether. The term freedom is the opportunity to be free. It does come with some rules! Our ancestors thousands of years ago would have had no such concept of the phrase freedom. But today, it is beholden to every living human, no matter where one lives. Embrace it. Do not take it for granted!

Human history has experienced hard lessons with much suffering to the masses when seeking freedom. There are dozens of free countries in the world, each with slightly varied expressions or limitations regarding the acts associated with what we label "freedom." No living creature chooses to live in a cage. Personal space, the choices one can make, and places one can travel are instinctual to animal needs. The necessity to find food, mates, space to dwell, and roam the lands of this earth are installed in its living creatures. Animals migrate, emigrate, coexist, and relocate based on innate forces. Humans too need to roam. It is the essence of freedom. Of all the beasts on this planet, humans have the greatest desire to not only explore this earth but also beyond into the heavens. Perhaps it is there where we find true freedom.

Some four to five hundred thousand years ago, during the time of *Homo heidelbergensis*, humans were hovering around a forged fire at night and began to wonder beyond the realities of their daily life. They relived the questions about their constant struggles to survive. There was no *"first day of school all over again."* Every day was new and different. They perhaps started the process of contemplating their importance and need for existence. These were not mindless human ancestors. They had developed skills to survive quite well for several hundred thousand years. Therefore, they experienced instead *"another day of school all over again."* It began with those distant ancestors continuing to learn and not make those same mistakes. It is something that we should not fear. When in error, learn!

From childhood experience, through times when my wife and I raised our own children, we watched and assisted in their development and maturity. Even now, with our grandchildren, the feelings and visions of events being repeated occur once again, and so frequently. Like the infamous saying, *"It's like déjà vu all over again."* Many times, I see these "happenings" as pleasures. Other times, I'll say, *"Oh, please God, not again."* We all go through that cognitive impression like it's *"the first day of school all over again."* The nervousness or anxieties from some of these events are just part of life. Perhaps it is some mystically related form of a cosmic joke to keep ourselves on our toes. Perhaps it is a way for us not to get too comfortable in our surroundings. In the animal world, the old, weak, or sick are taken out by nature. In human nature, we either adjust, or we inadvertently take ourselves out. Either way, nature has its course of action. It's up to us to somehow recognize it and accept our own consequences. Somehow, we seem to continue to go forward for a while and then backward once again. And so is the lesson of *"It's the first day of school all over again."*

Chapter 5: Inside the Mind of a Child

Inside the mind of a child, or through the eyes of a child, are limitless possibilities and images they can possess. It is almost impossible to have any discussion at all about human nature and human behavior without talking about family. And with family come the kids. The single greatest asset to human beings is the *family*. We generally think of family as our bloodline from one generation to the next. However, there is nothing wrong with the idea of discussing family as members who are not genetically related but are of great value to the family tribe. Yes, we are of one species, with the phrase "race" being completely inappropriate today. I don't mean this from a discriminating point of view, but from a biological perspective. People of this earth are too genetically alike to be separated into different races today. It is a slang term, and a rather insulting phrase to use now. We are of one race: the human race, and the last of our species. The removal of the phrase "race" could, or should, also be obliterated from a humanitarian viewpoint as well.

At this day and age, human anatomy and physiology are far too common compared to ancestral deviations, in addition to the fact that we cannot afford to segregate ourselves by a meaningless title. That is why it is vital as to how we treat ourselves and what we do to and for one another. To jeopardize humankind is to threaten our species to its final end. Nature is unforgiving; extinction does not allow for its return. Once a species becomes extinct, it is gone forever. Please do not ask about our futile ability to manipulate DNA to revive past species to dwell on earth again. On paper, it appears very simple. We have cloned some plants and animals in attempting to bring back extinct species or extend desired results within them. But should we? Nature has executed this act for a reason. Who are we to go against the natural effects of nature's plans? Who do we claim to be? Fortunately, for our young ones, the potential realities of our own species or unwittingly self-inflicting destruction do not meander inside the mind of a child.

As intelligent as humans perceive themselves, there are frequent incidents when we just don't get it. From an evolutionary point of view, our cranial size increased because of a large brain. This resulted in what our early ancestors began to eat. Changes in ancestral diets, from being plant eaters to scavengers and eventually hunters, led us to consume more protein from meat. Therefore, that increased muscle mass and brain size. Add the necessary minerals from foods eaten, and the synapse connections within the brain spread like wildfire. It is amazing that our opportunity to change diets increases our chances of surviving much better than our competitors. But a large brain does not directly reflect intelligence. You could own a library of books, but if you can't read, they're useless to you. So, intelligence and reasoning are more than just the hardware one possesses.

Evolution is all about change. It doesn't cross the boundaries of one's own spiritual beliefs or suggest how life began on earth. I personally see a connection between science and undefined spiritual influences in this subject. However, one can verify and quantify changes and advances based upon specific changes being at the right place at the right time. The detailed events that perfectly align may correspond to potential advantages. Is it all by design, chance, or luck, that combine genetic components and make us who we are? In any event, human nature is a mysterious puzzle at best. So, the questions may be, whose eyes do we see through to understand the human puzzle? I like to think that we should begin by looking through the eyes of a child. Children possess abilities that we seem to lose with age. And that is not necessarily a good thing. The mind of a child is golden. I'd like to think that some aspects of a child's mind still reside in us as adults. Perhaps then, prejudiced views, noncompromising actions, and selfish thoughts would not be so ubiquitous.

Human existence has dwelled on earth for an extremely short period of time considering the age of our planet being some 4.6 billion years old. Life as we recognize it began some five hundred-plus *million* years ago. Of course, the first living creatures were the simplest in structure and function, one-celled organisms; bacteria that go back to approximately 3.5 billion years. Sorry, no human creature was there at that time to verify the exact timeline, but scientific evidence in the way of fossils suggests otherwise. Fossil evidence, homologous structures among different creatures, embryological evidence, and now DNA sequencing, allow for the theory of evolution to exist. However, with every single scientific theory, it can change with additional science-based data. Afterwards, new theories arise. One of the greatest things in science is to be proven wrong. This encourages us to work harder toward an absolute truth. Hopefully, there are no egos in scientific endeavors.

There are still a ton of holes in people's concepts of evolution, but we can at least state that organisms do change over time and evolve. This may be from a little luck, the right mutation, unique adaptation, or an intervention yet to be announced. Does the truth lie with Darwin's views versus Gould and Eldridge's theories, or from Bob the bartender up the street and his beliefs and ideas? It almost doesn't matter. I can only discuss that which I currently know based on obtained information. Tomorrow, I hope to know even more than I do today, but won't lose any sleep over it. Everything changes: theories on evolution, the weather, and even stubborn minds may change.

As a retired science teacher, with 20 years in this field, and no insult to science educators, current science doesn't have, nor ever will have, the absolute answers to the origins of life. As for me, the real question about life is *why*, not necessarily *how* things began so many years ago. Perhaps that is the component in science which is not part of this disciplined, logical, systematic search on the subject. Still, it is human nature to wonder, speculate, and seek information to know where we came from. That being said, why don't we begin to seriously think about where we may be going? After all, doesn't that have the most impact on our daily lives? It is nice to know where we came from, but I'd like to know where we're going. And for that, my friends, you need to know a little bit about human nature and the behaviors that drive us. The most striking thing about human behavior is when an individual finally comprehends a complex subject. Their face lights up, that beautiful smile appears, and more importantly, they begin to feel good about themselves. It may not be verbally expressed when they become enlightened, but self-confidence may be the single most important attribute for success and survival. If you've got it, you can do anything.

Humans have a way of showing their resolve when they are determined to succeed. And it all begins with confidence and faith. Kids are the best at expressing their acknowledgment of understanding. Why, at times, do they somehow appear smarter than adults? You can see the delight through the eyes of a child. It may just be that at their young age, the eyes of a child are hypothetically more open to endless possibilities than their elders. It is suspected that our ancestral young ones had a limited contribution to the success of the tribe compared to those providing food and shelter. And it goes without saying that the Neanderthal moms were probably looked at as nothing more than resources to ensure the group's population. Thankfully, things have greatly changed over the last fifty thousand years.

However, being human has its unique behaviors, social structure, and order to endure the harsh realities of a cruel, hard world. At least today, we can inflict some humor upon others, as we often watch their futile regiment for pursuing those unobtainable personal goals. The human trait of humor evolved to become a common element of humankind.

*   *   *

Consider the times we took our kids to see Santa Claus before Christmas. They are everywhere. Department stores, garden centers, you name it—there is a Santa waiting for young faces to show up. When my young children were dragged to see him, they quickly figured out he could never be in all those locations at the same time. We smartly said, of course not, there are many Santa helpers, but you never know where the real one is sitting. My wife and I would eyeball the right Santa for the kids to see. Most wore a scruffy suit, food in their beard, and smelled like cabbage and cigarettes. Well, we would travel around to find what we considered the legit deal. One year, we chose a Santa who was located at a Christmas tree center that appeared very authentic. My young son, at a very early age, saw him, walked up without crying for once, sat on his lap without any bribery, and told Santa what he wanted for Christmas. As soon as he and his sister were finished, they rushed over to my wife and me and *whispered*, "I think he is the real one." Success! We pulled off the perfect Santa, without any tears, and their fantasy continued. What a wonderful life!

Life takes on a whole new perspective when daily humor and a little laid-back, less serious approach to living are implemented. I've seen some unique ways in which other people accomplish this. That "less serious approach to everyday living" is sometimes hard to achieve. But how serious should life be taken? Don't get me wrong, the things we do are important and can have drastic effects on ourselves and others. Respectful humor toward us and others is a wave of fresh air, which is much needed. We must cope with relentless chaotic stress from what people are forced to endure daily. Of course, many personal difficulties are a matter of perspective or maturity in how and when to place them in a box to be addressed at a later time. Human nature may choose to put them off, but that box will eventually need to be opened. Try a little humor and levity. Our ancestors didn't understand this, which is so unfortunate for them. But when do we actually define what it is to be human? Or have we been all along and changing definitions with the changing times? Now enter the conditioning mind of a child.

I have been intrigued and captivated by the mind of a child ever since my firstborn many years ago. It is difficult to understand the numerous images that they observe and perceive within their own world. As they try to understand their world, I could almost feel as though I was sucked into their little minds. I have interacted and shared my life with my two children when they were very young. Both of whom marveled at learning new things. The term *unique* does not describe their incredible ability to interpret life as they had experienced it. It is as if by magic that I, too, was transformed into their openly receptive, observably naïve, truly innocent, and unconditionally loving world. I treasure that gift from them more than anything else in the world. And being part of that shared experience, what I've received can never be taken away by anyone.

Animals can, in certain ways, simulate human behaviors. There are so many animals today that instinctively defend or offer themselves to predators to protect their young. An elephant will leave their herd and remain with its dying mate until the end comes and then eventually move on as if their sacrifice was to honor their now deceased lifelong partner. Or perhaps it is about a sense of love unknown to human love that keeps these creatures together. And there are hundreds more in the animal kingdom who display compassion for mates, which cannot be instinct alone. There are more feelings and higher-level thinking among animal kin than I believe science realizes. Although bonds exist between the young and parental animals, it is nearly impossible to completely understand emotional attachment between species of most animals.

Speaking as an individual and not as a biologist, I feel that there are a multitude of emotions that exist at various levels between animals that we'll never fully comprehend. And one main reason is the lack of complete conceptual language between them and us. That being said, I can say that I do see love in my golden retriever's eyes as they look at me with only their desire to seek nothing more than my return affection. My golden retrievers know how to play me. They may look innocent, but they know exactly what to do to get a small treat or large Milk Bone. The same is true when your kids give you those puppy dog eyes to get an ice cream from the Mister Softy ice cream truck. Who could resist those eyes, either from your dog or your kids?

As I had watched my own children grow and develop, it returned me to a time when I had childhood fantasies and illusions of a reality that worked for me. Contrary to what many believe, the mind of a child is not so simple and basic as it appears by their actions. There is more processing at an early age than what occurs in some adult brains. The complexities may differ, but the content is immense within the mind of a child. They are like walking, talking sponges, gathering and processing information at tremendous rates. Over time, they learn to develop the means to retrieve that data and apply it to the real world. This is what the learning curve is all about. And it is all achieved in different ways and often by unique techniques. It is a part of human nature installed for our very survival. Little do we realize that everything being gathered by our senses is processed, with some information being stored in specific areas of the brain to be retrieved later. Other stimuli are ignored and dumped as being perceived as having little value for the individual. This intricate process is no accident of evolution. It is a process that is too complex, and yet aspects of these physiological methods of obtaining, storing, and discarding information are similar among other lower-level primates other than humans.

So as adults, can we remember what it was like to see through the eyes of a child? Vision, being a primary sense, is utilized to evaluate and understand the world, often clouded or misdirected by what we visually perceive. But unlike most adults, children see more than just with their eyes. Perception or childhood vision often occurs in their minds as opposed to what we now, as adults, process through more than our sense of sight. We then relate the stimulus to past information for the correlation of what is right or wrong and safe or dangerous. Adults also process value in what we perceive. Children need to learn what is considered valuable. Adults believe they know what is important. Perhaps it is their hubris attitude that reinforces that. These lessons build their morality as they develop from the guidance of moral parenting. It is human nature to want to learn, just as it is human nature for a newborn calf to rise to its feet right after it is born.

Children also learn very quickly by imitation of what they witness. The expression of polite manners and respect toward others shown to them, the more they become incorporated into their daily character. Our ancestors' children thousands of years ago would have gained much of their knowledge through observation and imitation of adult members. Things haven't changed much. As their mental capacities expanded, reflection, reasoning, and judgment slowly became part of their learning process and behavior that would eventually be an advantage to them and their kin. Your actions and behaviors in front of your children are equally important to what you say to them, or, better yet, what they *hear* from you.

Children today also use the same approach toward moral and ethical behaviors as they mature. However, their incredibly complex network of neural synapses within their brains is more advanced than our ancestors', which required more evolutionary time. As interesting as the mind of a child is, what they convey and contemplate with their eyes is of most interest to me. What do they actually see or perceive that we as adults can no longer detect? And yes, I believe there are great differences in what we perceive as adults compared to what they see. I am not referring to simply understanding based on maturity or immaturity of neural processing. My statement about sight between the very young and us as adults is not about the anatomy or physiology of the eye. It is all about *"sight"* having a meaning all its own. It is having sight, but not the ability of *"vision."* As you may be aware, some never achieve this gift of inner sight.

And that brings us to the heart of children. How extraordinary are the views and perceptions as seen through the eyes of a child? Even more complex is what occurs inside the mind of a child. It is one of the last earthly mysteries and frontiers that we as adults may not have the ability to comprehend. Adults left that realm years ago with little chance of recalling how and what is being processed, especially in today's world. I am absolutely astounded as to how young children interpret and make sense of their world. Yes, their world, not ours.

I have had my own young children whom I've been very close to and now grandchildren who follow me like my shadow. When they're with me, we are inseparable, and I get to see how they respond to this confusing world. However, children do get confused, but with the right guidance, they will figure things out. Sometimes they're right, and other times they're not. Many adults, on the other hand, lack the patience to work through obstacles. How often do adults say, *"Oh, the hell with it?"* Humans often seek a quick answer. And because of this, we struggle for truth and security. For children, they'll seek whatever means necessary to make their life content. And I think that is wonderful. Perhaps it is we who should learn from them.

I will admit that I've learned a great deal from my own kids when they were young and now from my young grandkids. It is a gift that I cherish, especially since I recognize what they are offering me with their insight and perception. I sometimes miss my naïve views of life that I once had. Kids really are smart. They may not be as savvy as adults, but they don't need to be. Make it simple, that's what makes the most sense. They know more about Occam's razor (*the simplest reasonable explanation is often the best solution)* than most adults I know. No insult intended, but I think kids may know more practical things than their elders. Keep in mind that human nature in children has a built-in measure of self-indulgence. The reason is that over time, they may understand what the truth is and what has value, versus what is useless and fictional.

*　　　*　　　*

Kids take that which is apparent or what they perceive and convert it into terms that make sense to them. One way or another, we all want to understand whatever is in front of us. One day, many years ago, my five-year-old daughter wanted to wear her wrist bracelet that she had received as a gift for a particular occasion. Well, at her age, she couldn't quite remember the term "wrist." She asked for the jewelry that goes on her "arm ankle." Since she couldn't remember the term wrist, she referred to it the only way that was logical to her. My daughter related the ankle to her wrist. *Clever* girl. She knew that ankles and wrists were similar in function. So, she used her knowledge to convey what she was thinking about. It is almost like kids have a vision in mind, which we later call memory, but for them, association needs to take a logical pathway. Where an adult may use recall for obtaining information, children use visualization within their minds by gathering what and how they see the world. And as you can imagine, what we know is what we perceive.

Occasionally, whatever may appear obvious to us as adults is strangely obscure to kids. This may be due to their ability to visually see the facts as they are presented, but thinking forward to an eventual outcome is often elusive. Many years ago, my family, my wife's sister, and her family went to visit Niagara Falls on the Canadian side. It was a great three- or four-day trip. During our stay, my six-year-old son and I, along with my brother-in-law and his son, decided to take part in a projected simulated ride, which appeared to send us in a barrel over the Falls. It was designed to be very elaborate with an anticipated thrill of going over Niagara Falls in a barrel.

This was a simulated experience with its 3-D projection video screen. The gentlemen coordinating the simulation added the drama of going over the falls to enhance the experience. When the man had finished speaking to our group, my son slowly looked at me. He very quietly said, *"What if we don't survive?"* I didn't know whether to laugh or cry from his solemn, but very calm, dread of possibly not surviving the experience. So, I smiled reassuringly, hugged him, and said, *"You'll be with me, so we'll be fine."* From that day to the thousands of other experiences during those early years with my kids, I started to realize how little ones perceive the world. I became more insightful with my children, who had no rhyme or reason as to how they were educating me. It was either a combination of their intuitive nature to convey their feelings or my skills of observation and evaluation that I learned to greatly benefit from my children's lives. In either case, I am a better person because of them.

* * *

Our distant ancestors did not view their children as precious cargo as we do today. There was always value in keeping the tribe's population counts up even when basic resources like food were scarce. However, they witnessed that many members of the group did not live very long. And once they were deceased, they no longer had value. Motherly instinct is just that. It is an innate behavior for many female animals such as primates to especially care for their young's developmental growth to maturity. This natural emotion among our ancestors, I feel, was the same as it is today. There is also evidence of this among numerous other animals. Safety in numbers is true against predators but additional mouths to feed can be a huge burden. Without sounding cruel or disconcerting, people within the ancestral tribe many years ago under specific situations could either mean trouble for more mouths to feed or have a positive asset to help make the group that much stronger. And strength with a balanced number of individuals in a tribe just adds to their survival. Children started to become valued assets. This can be true particularly true with the hunt for food.

Children today love to explore both physically and mentally. It is a progression in learning and development. There are two ways in which human behavior flirts with danger. Literally, as in going out on a precarious limb with risk of physical injury or the dangers and risks that evolve within the mind as unsound ideas or intentions. The multitudes of influential media scenarios that depict risk as excitement, taint the young, underdeveloped reasoning in children in thinking that which is viewed in "game-mode" can in a way transcend into reality. As if it weren't difficult enough to differentiate what is played out for entertainment compared to how it could translate into reality situations. Risk must be asset and questioned to its purpose and value. Young children have some difficulty with this concept. Trial and error can be fatal.

I've been extremely fortunate that I didn't have any serious issues that occurred during my kid's adolescent years. Because of various forms of media stimulation, I do believe children today have a monumental hurdle to cross in segregating certain aspects between the two worlds of fantasy and reality. Whether one realizes it or not, it is mind conditioning. Add to the problem, as any biologist will tell you, is that the ability to make good or bad judgment calls is the last developmental stage in the human brain. The ability to make those sound decisions is not often reached until approximately twenty-five years of age. So, as a large complex society, we're basically screwed.

The first thing that one generally thinks about when it comes to children is, in a word, "play". The word itself defines fun, euphoria, excitement, thrill and mystery just to name a few adjectives. It allows one to search into the imagination of unexplored worlds. As a kid the world didn't revolve around the sun, it revolved around fun. And of course, fun and play go hand in hand. Even when I was a kid, which I won't tell you how many years ago that was, we played army against our buddies. All you needed to do was pick up a good stick and you had a gun. Your imagination was everything. It was a magical realm within an existing world which you had no clue about. Needless to say, my friends and I were closer to this side of being poor, so fancy toys didn't necessarily exist for us. We played with a baseball until the leather came off the ball. A game of football lasted until the neighborhood dog stole the ball. We'd chase after him for hours to get it back. Still, playing with a dog drool-soaking football did not phase us in the least. That's just the way it was.

Creativity and the ability to think beyond what was in front of us is an evolutionary development. In those early days everyone in the neighborhood had to be *inventive* because things weren't just handed to us by our parents; we creatively adjusted. Believe it or not, but we hand made our first hockey sticks with a piece of wood and section of eggshell floor molding as a blade. We didn't care or complain. We thought we were pretty damn smart. We used a small ball as a puck to play hockey in the street and made our own goalie nets. I was occasionally the goalie because you had to make your own protective padding. Nobody had a cup, so you shoved a big wad of foam rubber from a pillow down your pants to protect your jewels. And then suddenly, we were hockey players envisioning ourselves in the Stanley Cup playoffs. We all gave ourselves famous Canadian hockey player's names that we couldn't really pronounce. But we were heroes and great athletes in our own minds. Glorious days!

Growing up in days when "toys' were not readily available made us more appreciative of that which we had and gained over time. The amazing thing was that I never rationalized it at the time. We didn't wish that we had things which weren't at our disposable. We improvised, made do, or did without with no resentment. Life *conditioned* us that way. Years later I can see that it helped me in the manner which created what I am today. My friends and I became resourceful at an early age. I think that it made us better people overall. We didn't think about it, but we appreciated what we had. Kids don't often think at such high levels of reflection, but it sinks in and is recognized much later. As long ago as it's been, it feels as though it was only yesterday. These are the precious gifts of memories which last forever. They really are priceless, and the most valuable thing is, you own. I would not change one thing in my life, the sorrow nor the joy.

I sometimes wish the eyes of a child could be captured in a photo. They are now images that remain fondly in my mind. I remember each of my many young friends dearly. I love them as brothers. But as children, we often fail to tell them how special they are and significant they have come to be for us. As children we don't always know the need to express our true feelings to friends that have influenced and molded our lives forever. The best we can do is tell them at later dates or if lucky in prose that they will hopefully read. I am the most fortunate person to have had childhood friends who have helped make me what I am today. That childhood bonding and personal growth made by all of us is a part of human nature. It often goes unseen and sometimes details are forgotten but at least should occasionally be remembered with the tales of the past shared with others.

* * *

Since this chapter is titled *Inside the Mind of a Child,* I can write about myself and my own children to some degree. It's impossible to write about others. However, we all have events that can be found on common ground. The fact that I am quite a bit older now does not mean that my mind longs for the past. I simply enjoy recollecting all the wonderful memories and wishes that were part of my childhood. I'm still probably somewhat of a kid at heart, and I hope that never diminishes. I do unleash that inner child at various moments to the dismay of my wife. Nonetheless, I'd rather be considered a child sometimes than an adult asshole.

I'm not an authority on human history to comment on human behaviors over the last two or three thousand years regarding play and fun activities. Nor can I explain the numerous idiosyncrasies that label human entertaining behavior. I'm not really that interested. However, in today's world, we must be cautious as to the amount of time we dedicate to each of those kinds of pleasures that we seek.

Perhaps a concerning thought is how we approach this topic. Hopefully, we will still maintain integrity and respect through the actions we conduct in the ways that humans have their fun. At the very least, it must be respectful and bring no harm to anyone.

* * *

Comical memories are priceless. Some of them come from kids through their acts of fun and play. You can recall them at your leisure, although there may be people wondering why you're smiling so often. But they don't really need to know why. I have various forms of physical workouts and training routines in my life. One serious and personally meaningful discipline that I undertake is training with a samurai sword in the privacy of my backyard. My young grandson was fascinated by the rhythmic and graceful movements of the routine I practiced.

When he was around three years of age, my grandson too wanted to *"practice swords."* So, being willing to bring him into the appreciation of this skill, I purchased him nonlethal practice swords for play more than anything else. He had several plastic weapons, including foam rubber swords, pirate swords, and anything that resembled a saber. As clever as we believe we are, we're always taken by surprise by what is inside the mind of a child. This is because whatever goes through the eyes of a child can completely fool us.

One morning, while he was at my home, I dressed accordingly to work out with my sword. He too put his little outfit on, and I wrapped a karate white belt around his waist about three times because he was so small. He then proceeded to acquire his weapons for practice. As he came out of the Florida room door into the backyard where I was, tucked into his belt was a plastic samurai sword, a pirate sword, a pirate pistol, and a plastic machine gun. He looked like a miniature Arnold Schwarzenegger ready to do battle against an army of terrorists. I had no words. For the first time, I was speechless. I eventually told him that he looked *fantastic*. The greatest thing was that he felt wonderful about himself. There was no way he could have shoved another weapon under his belt. In his mind, he was the ultimate samurai. That's when I was reminded that life was good.

Comic genius isn't created; it is born within certain individuals. As adults, we seem to know what is hilarious and funny for our personalities. But children are innately gifted with a sense of humor that they don't necessarily recognize. Call their young behaviors quirkiness or unusual, but through their eyes, they act as perfectly normal humans. Perhaps it is because of their young developing brains. But intellectually, humor and that which is comical are two different things. Kids are naturally funny even when they don't intend to be. Innocence may have much to do with this. And that can be a horrible thing to lose. Enjoy it when you see it!

Now, my son, when he was three or four years of age, would always dress himself by wearing his PJ's, authentic Indiana Jones fedora, and snow boots throughout the house. Even though wearing snow boots in the house wasn't an expectation of snow, my son simply enjoyed it. He was obsessed with Indiana Jones. That is exactly who he wanted to be when he was all grown up. He wasn't expecting snow to fall from the living room ceiling but wore his snow boots each day, even in the summer. My wife and I would get strange looks from adults while we walked through the garden center during summer with our kids, and my son wore his snow boots. Of course, they were always superhero snow boots. And I have to admit, I wish I had a pair myself.

* * *

The best we can do to see through the eyes of a child is pay attention to our own young ones. It is a time, short-lived. To allow it to go by unnoticed would be unforgiving to any parent. We may be able to witness funny and sometimes unusual behaviors of our kids, but we'll never fully understand what is inside the mind of a child. That is a secret private realm which they alone should be privy to. Nonetheless, I find it sad that much of those characteristics fade with time. At least we have the memories and family tales to share about our unique past. It would be a shame not to pass along those moments of precious treasures from generation to generation.

Chapter 6: The Shape of Shadows / New Version of an Old Story

The funny thing about humans is that we believe whatever is perceived by using our given senses. As everyone knows, our five senses can often mislead us from a presumed truth. Our brain interprets input from the senses and sorts through information to determine logical sequences of whatever we perceive to be real. The mind also likes to force our brains to dish out safe, comfortable, non-threatening information that may keep us free from danger. On the other hand, the mind and brain work together in assisting us to escape peril. There must be a constant battle going on in our heads. No wonder we're always confused or frustrated about life around us. It is certainly pointless to question everything that we see, smell, or touch. So, there is an unconscious point during development when we tell ourselves, OK, I trust what that little man in my brain is telling me.

I've always said that perception is reality. I know that doesn't sound scientific or logical. But I do understand from a physiological viewpoint how the senses can be fooled. Add the complexities of the emotions and mysteries of the mind and brain, and we have little chance of being correct only half the time. I don't wish to be too philosophical about reality, but people are easily fooled by no fault of their own. For kids, reality has little value as they gravitate to the pleasures of imagination, wonder, and mystery. Unfortunately for most, these cool abilities dissipate with age. It is another reason why I chose to retain some of my childhood vices. They're fun, and who wants to give up their childhood collection of imaginative treasures? So, it is either that or my incredible charm that my kids and grandchildren gravitate to my personality. It will be a sad day, if or when, I ever actually grow up and lose that which I have held so dearly for years. This is not to say that I can't undertake the highest demands of being a responsible, productive, and effective adult. With all of my positive and negative qualities, I'm still only human.

I find it incredibly humorous to see some people who adorn themselves in fancy outerwear, or more like costumes, to express and imagine themselves as a character they've chosen. Expression is fine, in fact I encourage it. Halloween is the only one day out of the year we can do this. But you can't be something that you're really not. Have fun with it. All guys subconsciously want to dress up as Elvis at least once. But I think my school administrator would have been a little upset if every day I walked into teaching my biology class dressed as Elvis. However, there are those who choose to pursue and live out a persona which is a little over the top. This actually goes back to early ancestral days of showing dominance by displaying tribal leadership. Expression is still a quality of human behavior which is perfectly acceptable. My suggestion, let's keep our actions for the right reasons. It does make me wonder about the first tribal member to think of shoving a bone through his or her nose and thinking that this would be an attractive look.

The shape of shadows resembles something potentially tangible that is part of a real object cast by light. They are reflections without becoming an exact mirror-detailed image. But we can often recognize the subject by its shadow silhouette. Movements mimic the motion of the object. Little kids have fun trying to step on their own shadows once they recognize them. They will attempt to emulate the motion of their shadows, not realizing they're the result of every cast shadow. We joke with them about "don't jump on my shadow" or go ahead and try, as we quickly move away in their attempt. For a young kid, shadows are mysterious and unclear as to their origin. But shadows bring no fear to the mind. They may be an unknown mystery to them but are harmless. Nonetheless, they quickly understand that shadows will follow them wherever they go. And I think they like it. Subconsciously, we may see our actions as being followed and watched by another.

Shadows cast at night in a dimly lit bedroom, for kids, take on a whole different perspective. The vagueness in identifying these shadows can be confusing and does not fit into their precise cubicle of reality. Without placing information gathered by the senses into known perspectives, it stimulates fear, even in adults. Only then do shadows take on a whole new meaning of what they are and why they are present. Are they friend or foe? This is the only way in which the mind is attempting to process information for the sake of our safety. It's an innate quality thankfully installed in human behavior. Fortunately, shadows do not have a mind of their own. And the mystery which surrounds them can be entertaining or terrifying.

For many, as we mature our ambitions grow and that desire to search for the one thing which identifies us becomes clear, and shadows take on a new form. Then they begin to form in our minds. Cognitively, we reflect on images or visions which may resemble past shadows to formulate ideas on how to proceed down paths which we've chosen. These shadows then become reoccurring in our minds in the attempt to help us sort things out or justify an event. They become more than a memory or recalling of information. These non-tangible, non-traditional shadows are a form of psychological guidance which doesn't exactly fall into any logical category. It may come close to gut feelings or instinct. Without getting into an argument over semantics, the "shadow" goes beyond that of a memory or storage of past information. And what emerges from many shadows of the mind is a clearly defined image of a personal identity.

Our ancestors struggled from the time when their intellect enabled them to contemplate who they were and what their purpose was here in this life. It became more than the need to survive. Human nature now drives us to become more than our capabilities to make it through this harsh world. We need purpose and fulfillment. It is one of many factors that make us human. To our current knowledge, we are the only animal species that feels the need to be successful. Whether this is to be loved by our own kind or to fulfill our hopes and dreams. Our shadow now becomes more than an image. It is a representation that walks on this earth not to dominate other life forms but to express our inner being. For we now search for more than how we appear. We strive for what we must become.

Human nature, at the very least, is a bit of a mystery. It is also true about human behavior. Sometimes we just don't get it. Other times we believe that we do. As intelligent as people can be, there's always an enigmatic presence within the world that defies explanation. Many reading this will say, "Well, we just need more time to figure it out, or there is a logical scientific explanation for all events and circumstances." Maybe, but perhaps some mysteries don't fall into the categories of the explainable or play by the rules of the laws of physics. Humans are only as smart as the tools we're capable of using. We are also as smart as the information we can recall. And lastly, humans are only as smart as our ability to execute such knowledge. Some may believe that humans are capable of learning anything if given enough time. Pretty hubris if you ask me. There will always be something that we don't know and will consistently remain unknown. Some mysteries in life will forever remain just that. Perhaps we should look at this as a dangling carrot to always attempt to improve our species.

There are people who like to thrive on mystical phenomena and others who do not. Some can disprove hypotheses and theories, and others love conspiracy theories. Remember what I said about reality and perception. Science certainly denies the concept of magic, but some mysteries are magical by the way they fool the senses or conflict with our concept of what is real. I would like to believe that part of this mystery tour that we travel through human evolution keeps us interested enough to continue exploration for the improvement of understanding humankind and the world around us. Let us not permit the nature of humans to be dominated by hubris behavior which makes us appear arrogant. Besides, we can always have a little fun with an enigma or the unknown. There are some things in life that we just need to leave alone. Therefore, leave the mystery and magic to those who enjoy the concept among the games of the mind.

Mystery and shadow can go hand in hand. Vague images of various subjects which are unclear haunt our consciousness all the time. We believe that our intellect should solve that which we temporarily misunderstand. But human behavior is not always taken as seriously as some expect it should. Enter the pagan holiday of Halloween.

Dressing in costume is a novel character plot during Halloween. It is entertaining. This yearly ritual purposely goes against the norm and creates shadows which normally would not exist. It gives us a brief moment to be someone unlike that in which the hand we're dealt. For many individuals, including myself, playing a secretive inner role model during the Halloween season is enjoyable. I have even stretched that behavior ever so slightly in the guise of amusement. I won't say how many times that when asked my identity I've innocently responded, "James, James Bond." Sometimes you get an odd look, and other times it goes unregistered by the recipient. Well, the first name is correct for me, just not the surname. It's certainly more difficult to play out the persona of James Bond, but what real man wouldn't want to?

You must admit that James Bond is an absolute classic among respectable characters. He saves the world. He's smart, good-looking, has all the greatest tools and weapons at his disposal, is highly skilled in beating and outwitting bad guys, and lastly is attractive to all women. Sure, I'll take that.

Things have changed from earlier times when our ancestors were happy to have caught and killed an animal for a good meal compared to today and acting out all of the suave acrobatic moves of James Bond. The more we progress as a species, the more we need to find unique personas that fit our inner desires or ways to be recognized. Hopefully the right role models will prevail. In my repertoire of respectable characters, I have more than just James Bond to emulate. But I suppose that's just the kid that occasionally comes out in me. What would we do without Halloween to spark such interest? It makes one think if we really are who we believe we are.

A new version of an old story may not be such a bad thing. Heroes and those who imitate heroes as children all have the same theme. It is generally a mission of good battling evil. And because humans innately have a moral consciousness and affinity for good over evil, heroes are born and live through all children's minds in reenacting a playful venture. Sometimes we seek out the unwanted hero which casts more mystery than hope. Who wouldn't like being that swashbuckling pirate? Feared by many, respected by none. It is a role that enables one to display oneself as an unknown quality to others. It is often better to play the role than to be the role.

Children play superheroes every day. They may even dress the part of their action figures as they run through the house, sword in hand, slaying dragons or demons. However, we don't concern ourselves when they wish to dress during Halloween as a ninja assassin. Actually, ninjas are assassins, so it's rather redundant. Quiet in the morning, then out emerges the bad kid of the hero-slaying dragon fighter an hour later. Human behavior today seems to have the need to display or act out that which we envision in our minds. And that indicates leaps of change from a millennium ago in our evolution. As we grow from childhood to adulthood, our inner shadow develops a clearer form which eventually defines who we really are. However, it must be a progressive sequence of events within human nature that takes us between being heroes and villains, to performing acts of bravery and becoming fearless saviors of others before we reach our end goal.

* * *

Many years ago, when my son was about five or six, he wanted to dress up as a Teenage Mutant Ninja Turtle for Halloween. Of course he wanted me to be the same. I think I was more excited about the idea than he was. My wife purchased all the materials for the costumes since she was the clever craftsperson who manufactured our outfits. She hand sewn fabric to make us look exactly like the original Ninja Turtles. My son loved watching them on television and playing with his action figures. We each had protruding turtle shells on our backs when we were dressed. He and I were mirror images of Teenage Mutant Ninja Turtles. We agreed on who was going to portray Leonardo, Donatello, Michelangelo, or Raphael. The next important thing was to acquire our weapons that were unique to each Ninja Turtle. Of course I chose the sword. It is kind of funny how I didn't get arrested that night. But if you're in costume, you must play the whole part.

It's Halloween and we were both set to go. We felt fantastic about our costumes and really dove into the characters. We were the real thing! Who cared about the candy? He and I roamed the streets of our neighborhood, a five year old kid and a fully grown man dressed as ninja assassins. We did our Mutant Ninja Turtle strut, eyeing up all of the other trick or treaters. He and I were ready for any possible threats to innocent civilians. Regardless of the strange glances and confusing looks at a grown man in a Ninja costume from other parents walking their children, it didn't matter. We were ready for action. We had the swagger and the appearance of power. The only disappointing thing that night was that we didn't get to beat up any bad guys. I'm sure when they saw Teenage Mutant Ninja Turtles strolling down the street, all the evil doers fled. It was a great memorable night for my son and me. Being slightly embarrassed now, I kept my costume on for days. I even wanted to wear it to work, but my wife ruined my enthusiasm. She said the co workers would only be jealous. I think she was just saying that to try to make me feel good.

It is so true that life repeats itself with slight twists in the specific details of events. Just a year ago my grandson, by his own choice, dressed as a ninja warrior for Halloween. It made me once again think back to when my son and I did something very similar. It is amazing how people can, by accident or design, be placed on the same page. How can we think or act in similar ways? It all has to do with human nature. As things change, they also seem to remain the same. The shadows of the past follow us and become part of our kin. It may be a part of the telling of tales or in the way we look at our past and like what we see. Everyone talks about and seeks change. And change is a good thing. Perhaps our own innate nature installs a synergy between remaining the same and embracing change.

*   *   *

Humans pride themselves on knowing the answers to difficult questions. From this, we believe we can develop an answer to anything. Perhaps we're asking the wrong questions. Then again, maybe we don't even need to ask. Who knows if certain aspects of our nature may need to remain constant? It just may be that it is temporarily elusive to us as humans. In any case, shadows will follow.

The shape of shadows has not really changed over the years. We evolve and the physical world changes. Yet visions in our minds of hopes, desires, ambitions, and expectations seem to remain the same. New versions of an old story surface in ways where people believe these new ideas and methods are unique or innovative. Some use this technique for personal profit, while others wish to be recognized as talented thinkers or avant garde saviors for a needy world. Let's hope that an original thought is not like an original sin. They each have origins in stories of the past long before today's world. Yet we throw around the term "original" almost carelessly.

There was a beginning for all things, but it has altered in form through the years. Shadows that we once thought to be comforting sometimes become confusing or misunderstood. This may delve into those aspects of our past that we have temporarily forgotten or placed on the back burner, hoping that they fade away or shift to serve a useful purpose. As difficult as it is to sometimes let go, it is also in our nature to force change for our future.

When it comes to human nature, and especially behaviors, we are who we are. Some things cannot be changed by force. I believe people can change, but I've never seen it happen. There is no denying who you are. Shadows occasionally return, often at unexpected moments. The second time may very well be in a new version of an old story.

* * *

Choice is a prime example of a shadow which we choose to create. Again, some of the roles we play during the festivities of Halloween have their roots in what our character or nature really is. Even as a kid I always wanted to be a pirate. Whether this stems from some form of rebellion or not liking being told what to do and when to do it, pirates were independent, aggressive, risk taking adventurers without fear. The downside is they were thieves. That truly is not me. I don't believe in pillaging and stealing from others, but the lore of a pirate's adventure on the high seas is enticing.

I've felt like this long before Disney's film version of **Pirates of the Caribbean** and its sequels came to the big screen. Who really doesn't love Johnny Depp's character as Captain Jack Sparrow? Sure, he's a "bad boy," but he does have a conscience and a moral code toward those whom he befriends. I may not actually resemble myself as Captain Jack Sparrow, but I do fancy myself as him occasionally during

Halloween. The reality is, my identity is *definitely* more like Orlando Bloom's character, Will Turner. He's bold, ambitious, determined, skilled, good looking, and of course gets the girl in the end. Nonetheless, it is fun to temporarily lose one's identity and to simulate another's for the amusement of our mind's fantasies.

I have no problem bending the reality of true historic pirates. They were ruthless and cruel in their day. But thanks to Disney, they've become somewhat of a folklore to kids, and of course to me. Human behavior seems to poke more fun at us today. We either laugh at ourselves or someone else is laughing at us. It doesn't matter as long as people are finding harmless amusement. It certainly softens reality since we all appear to need it.

If your mind spends too much time every day in the real world, you may wish to try finding ways to free your inner self. You might be smiling just a little more. The world absolutely needs that. How nice would it be waking up in the morning thinking, who shall I be today, Captain Jack Sparrow, Will Turner, a Teenage Mutant Ninja Turtle, or James Bond? Sadly, I put my big boy pants on and off I go. But no one knows what's really going on in my mind.

* * *

Taking on a false persona seems to become the norm while interacting among the masses. It's OK to have fun, but when it crosses that thin line between who you are and what you wish to be, it leaves an identity vacancy that the mind has difficulty adjusting to. It is a skill which must be mastered. Ask any true professional actor who plays many roles in films or theater. They can turn on a character for themselves to portray like turning on a light switch. However, this is not the norm for all people.

Our human nature pushes us not only to seek our true identity but also to be confident in undertaking a difficult task. Think of it as camouflage or like a human chameleon. The ability to bounce outside of our character during threatening or risky environments can save one's life. But this is quite different from role playing. Humans have innate behavior for survival tactics. So under the right circumstances we may not appear to be who we are as viewed by others during our ordinary lives. The body can withstand almost anything; it is the mind which needs training and discipline.

Many animal species mimic other species or display themselves differently to resemble a greater threat to opponents or predators in the wild. Wow, how much are they like people? Although fascinated by dragonflies, one of the most interesting insects to me is the "stick bug." Here is a harmless creature which nature chooses to survive by mimicking a twig or stick. It would be overlooked by almost any predator.

The viceroy butterfly physically mimics its wings by coloration patterns to that of a monarch butterfly. Why? Because apparently monarchs don't taste very good according to birds and other creatures that may want to eat butterflies. Go ahead and ask any bird. Therefore, the viceroy butterfly goes about its own business saying, "Nah, nah, you're not going to eat me." There are hundreds of animal species in which nature has provided them with potential sanctuary to continue onward by hiding in plain sight. Their shadows therefore never change. They are what they are and there is no need to change.

Today we have so many new versions of an old story that it's becoming monotonous and emotionally draining. You would think that after eleven or twelve thousand years of human development we would have progressed much further socially and morally. Of course, the term "progress" needs to be defined, which I'll leave to your interpretation. I say this because, in my mind, humans really started to advance as a population during the "Birth of Agriculture" period which emerged in the Middle East ten to twelve thousand years ago.

The human population at that time is estimated at approximately one million people, which occupied current Europe across Asia and into western North America. Shortly afterwards there is a theory of a western expedition from Europe hop skipping across a northern trail by sea and ice sheets to North America using Iceland, Greenland, and eastern Canadian regions. They didn't create shadows of illusion; shadows followed them. History is marked by those proposed shadows.

Our ancestors needed to overcome so many physical, emotional, and mental obstacles to survive. They did so out of a need by having determination, strength, and exploration instincts. We cannot rule out a simple desire to move forward and leave that which they came from. There would have been very little of any "new version of an old story" with our ancestors. It was either progress, adapt, or die. The shape of shadows was very clear and distinct for them.

Fortunately, our ancestors were perfecting that which was learned and passed along to offspring. They were exploring new ideas and ways to survive. There would have been little value to keep things exactly the same in a changing environment. Their human nature and behaviors would have been inventive and inquisitive, as it remains today. However, I believe they would have determined what actually worked and what did not. This would have been through many trials and errors. We of course refer to this as a learning curve.

So how much time do we as a species say, "OK, that's enough? Let's just drop this and move on." And why are there so many learning curves that are incredibly difficult to accomplish? In our current trials and errors to advance and succeed, is it learning curves, or the means to accept that which cannot be accomplished today? Patience, listening to others, and never giving up should be considered when approaching any learning curve.

* * *

If someone gave you a new unnamed board game with loads of fancy pieces, a mysterious unclear path along its playing board, mystery cards that give vague cues or no direction as to how to continue in the game, with rules that don't apply to any move along the game, and no clear purpose of how to win the game, you'd probably think that this was useless and a waste of time to play. If given to me, I'd think it was a piece of crap and throw it out. The question is, did you rule out the failures in the tasks of the game and seek any real objective, or just give up without using your personally gained knowledge to figure it out?

Life can sometimes resemble a game like this. That's because life is sometimes *crap*. But you don't throw it away. Instead, you spend a lifetime trying to figure out how to play it. Past shadows can enlighten new thought processes to accomplish that which currently eludes you. Don't give up the game!

Sometimes shadows can show direction toward a path we should follow. Other times those shadows are elusive and suddenly vanish. Shadows of enlightenment can lead to success if they are meant to be followed. Success is not by pure luck alone. Look at all of the notable entertainers, athletes, businessmen and women, and a multitude of others who have reached success by following their shadows, using their knowledge, taking risk, and being determined. But it does not happen to everyone. Many will fail more than once or twice, or seemingly always.

On occasion we need to slow down, take a breath, and regroup before moving forward. Everyone wants to win or be a winner. And if it doesn't work out the way you want, so be it. I tell people all the time, don't think of yourself as being a loser, you're just a non winner. Either way, to win or lose is personal, but no one should be labeled otherwise. The most important thing is not to give up on the game. I am not saying that we should take life carelessly, just maybe not so seriously. Just as Glenn Fry of the music band Eagles sings, "*Take it Easy.*" One of the best bands ever! {Of course you can't include the Beatles in that competition; they are the Fab Four for several reasons}. Perhaps we should take that "easy" approach to life. After all, it is challenging enough. Whatever is meant to be, will be.

As clever as humans believe they are, we've always had difficulty in recognizing that which we accept as truth. Perhaps it is because truth is personal and may differ from person to person. How many times do we see a new version of an old story? Those new versions do not always intend to be truthful. Remember people have their own agendas. And those agendas often are focused on oneself. Human behavior relies on absolutes for comfort reasons. People will question everything, and maybe that is a good thing. However, being critical and questionable about most people will not refresh your mind.

Trust is a non tangible quality. Sometimes you just need to have faith in yourself and others. If one group believes something to be true, another may disagree. Those shadows of what we think are real may be true. On the other hand, perhaps we need a clearer image of those shadows changing shape to become a truth. Are humans sometimes playing *a game with no rules, with no direction, and not making any sense*? YES, of course! You must figure it out. Humans have fought and died for differences in beliefs even when no possessions or artifacts are involved. Beliefs are highly personal and strongly defended. At times they become part of human culture and other times a phase which only momentarily shows itself. They become a shifting shadow as if blown by the wind. I never said that the shapes of shadows are always recognizable. Maybe to play life's game, you need a pair of loaded dice.

They say that playing professional baseball and hitting a ball is one of the hardest things to do among any sport. It is even harder for batters during a day game when evening is approaching and shadows are cast onto the field. There is tremendous difficulty in visually picking up the spin of the ball and its location when thrown by the pitcher. During that transition between daylight and the progressing evening, shadows on the field make the game even harder. This is especially so for hitters until those shadows dissipate and ambient lighting balances out between the contrast of dark and light reflections coming off the ball. Good professional ball players learn how to adjust to these physical changes, including gusts of wind, bright sunlight, or the falling of rain before a game is put on hold on account of weather. From my perspective, life is like baseball!

* * *

A new version of an old story occurred every time I'd attempt to teach my two kids, and eventually my grandkids, how to play baseball. Being a high school science teacher at that time and having most of the summer off, my daughter would drop off the grandkids while she worked during the day. My wife and I were very happy to watch them.

When my older grandson was about four years old, he was learning the sport of baseball. His brother was born several years later. I'd take the older grandson out to the front lawn and practice hitting plastic wiffle balls. I'd teach him how to stand at the plate, how to swing the bat, which pitch to swing at, and how to run the bases. It was almost a repeat of when my own son was his age. Life seems to rotate in a circle, and that's a good thing. There are always new versions of an old story.

There was an approximately six year difference between my grandsons. From day one they were as close emotionally together as could be, and that bond remains today. When the younger brother became of the appropriate age, I would also begin to teach him the sport of baseball. Therefore, as a proud pop pop, I would show my four year old grandson how to stand at the plate and swing at the perfect pitches. In the very early days, it took time to come in contact with a plastic ball. I'm sure it appeared to be coming at him at ninety five miles per hour. He was only four years old. I must admit he did well and learned quickly. The baseball field was our front lawn.

As I pitched the wiffle ball, I'd say, "No, no, outside, don't swing." And then it was, "No, no, inside," or "Too high." He'd swing and miss, swing and miss, and it turned into us just playing a game of catch. He would throw all the balls back to me and it was Deja Vu all over again. In no time at all he got the hang of it, especially as his dad also practiced with him when he came home from work. Soon he imagined himself as an All Star, with confidence growing each day.

We would practice often. During one perfect pitch right down the center of the plate, he cracked a line drive so fierce that it took a beeline right to my giblets. It was so quick that I couldn't even flinch to move out of the way. Down I go, bent in half, screaming at the ground in pain from a direct shot to the jewels. Fortunately it was a plastic wiffle ball, but every man knows that any kind of projectile object striking the family goods is going to cause unimaginable pain.

My grandson ran over to me wondering what had happened. So being bent over as I was, I explained how disappointed I'd been to see how much crabgrass was on the front lawn. It was the best excuse I could come up with. Eventually I straightened up to pitch the remaining part of our practice. The funny thing is that the very next day we had to start all over again with the technique of hitting baseballs. Another new version of an old story. We move forward and we move backward from time to time, and that's OK. But for me, I did learn to cover up myself after pitching.

Life can resemble some struggles like playing a game of ball with many influencing conditions. Like a ball player, to live this life we must adapt to huge or subtle changes to win. Unfortunately, the need may occur quicker than we are capable of making those changes. Those people who won't adjust, or can't, just fall to the side. Individuals who succeed will pass on the secrets they've learned along with the best way to play life's game. As a result, so many people are reliving a new version of an old story. Some may recognize their shadows, while others will never find them. I sometimes wonder how often we need to repeat our actions before behavior moves toward a direction away from old habits. Then again, some lessons are never really learned. And so is the shape of shadows.

Chapter 7: The Eternal Affection of a Dog

The history and domestication of a dog is as significant to their species as it is to humankind. Humans would not be the people we are without their presence. It was destiny that dogs and humans join emotionally as we have over the last several thousand years. We have become part of their group as living creatures just as they now belong in ours.

The relationship between adults and their children is an essential connection to a family's natural course of human nature. Dogs have become part of the human family. People are designed to be among a tribe, pack, or group. Our species was intended to live as a whole. The interactive behavior among the family is critical to its survival. We need each other. The domesticated dog, being of a different species, has also become part of the human family's unifying belonging, personal satisfaction, and expression of love.

Human nature dictates that we care about and nurture our own kind. Women carry their babies for nine months. Both male and female partners tend and protect the newborn child's every need. We also make every attempt to raise the child to be successful and content. Parents develop eternal emotional bonds. Those connections never truly break, no matter what type of fallout occurs through difficult times.

Humans are not solitary creatures. We need family and friends. It is an innate impulse to establish this.

Our development and behavior have evolved to also include creatures such as dogs to become part of our tightly bonded family. We could come up with several reasons for this, but there is something mystical and unknown as to why dogs, of all species, have the greatest effect on some people's emotions. They seek us to be part of their complete life.

The domesticated dog originated some 7,000 to 20,000 years ago, most probably in the southern East Asia region. The timeline is still disputed, but for me it doesn't matter. Dogs were conditioned to be tamed by humans. They played a significant role for us for the last five thousand years or longer. The scientific name for the domesticated dog is *Canis lupus familiaris.* It is unclear whether they are the descendants of the gray wolf, *Canis lupus,* or evolved as an offshoot directly from several different groups of wolves.

There are various species of wolves and other similar four legged creatures such as coyotes that are thought to be a cross between the gray and red wolf. It doesn't really matter how dogs evolved from wolves. They are descendants of them. However, there was a process to bring these animals into human populations. They were tamed and coexisted with early human hunter gatherers several thousand years ago. It was recognized early on by our ancestors that dogs had great value in many ways to the tribe. Perhaps they assisted in gathering animals to be hunted, defended their masters, or provided social pleasure among the tribe.

Obviously today we have many different breeds of dogs for pets, or rather members of the family. "Selective breeding" is a human method of discovery for crossing organisms to achieve a set of desired traits between two similar groups with varying characteristics. Genetic variation among any species is crucial for healthy survival.

For example, the cheetah today is so genetically identical among the members of their limited population. The inbreeding has resulted in risks that the slightest genetic mutation could run through the entire species, placing them in grave danger. This reduces population numbers to near extinction. Among all cheetahs living today there are only a few genetic differences. *Diversity ensures stability.* Changes created by human and natural influences have affected the cheetah to a threatening level of endangerment. Cheetahs are becoming closer to extinction because of this and reduced natural territories. They have a special place in my heart for their very existence and how we have affected their diminished population over recent years. Their numbers in the wild dwindle. There is *no* return from extinction.

Overbreeding in any one type of animal, including dogs, can exaggerate flaws in the offspring with less desirable characteristics. These flaws can be physical or behavioral ones. It has occurred with many breeds of dogs over the years simply due to people's demand to have them. Various undesirable physical and temperamental characteristics have placed certain breeds of dogs as becoming problematic to owners. Purebred dogs typically have more issues than a mutt or mixed breed. Slight genetic differences can be a good thing.

That being said, the technique and skill of training and raising dogs is so important for retaining a disposition or personality which is acceptable to its owner. People want what they want.

*   *   *

Animals too call out to life, as exemplified in the uncanny ability of our family's three golden retrievers to demand their supper. Sure enough, every day at 5 pm all three dogs sit very patiently and stare intently at me as I sit watching the news on television. It's a shame they can't speak our language to ask me to feed them their dinner. But their piercing eyes, fixed on every movement I make, clearly express their wishes for me to get off my ass and get them their supper. It may also include my wife if she is in the room. I swear they can read the clock to prompt me or my wife to get on with our feeding duties. This daily ritual occurs like a precise Swiss timepiece that I still find amazing today.

Professionals might argue that this behavior is learned and reinforced over time, for which I understand and agree. But I've noticed other behaviors in my four legged companions that suggest that they seek out our contact through needs and most likely through their desires. There must also be a running timepiece in their minds.

My wife and I have always had purebred golden retrievers as additional members of our family. Their personalities are all different, yet still similar, but the bottom line is that they're all great dogs and close companions to all family members. As domesticated and tame as they are, they still display a hierarchy of structure among themselves. Of the three golden retrievers that we currently have, the youngest believes that he can get away with anything and doesn't think twice about taking various toys away from the other two. He's the first to squeeze his way into getting a snack and thinks that he is the only one who should go for a walk.

The old girl will throw her weight around if she needs to, but she's the boss, or rather the queen of the house. When she greets people coming into our home, she'll immediately roll over for a belly rub. The middle child is the gentlest creature you could ever meet. It is amazing that you can love each of your dogs equally and differently, as you would your own children. No one is better than the other. And that makes them happy to be around, just like your children and grandchildren.

Golden retriever number two, Indiana, named after the film character Indiana Jones and nicknamed "Indy," will lay at my feet and insist on placing his paw on my foot as I sit in my family room chair watching television. If I move my foot, he'll search for it with his paw until he finds it, as if to make constant contact with me. Our young one insists on sitting his eighty five pound body mass on my lap as I'm in my chair just to get his affectionate petting and loving. Believing that he's still a small cute little puppy, he really doesn't have a clue about how big he is.

The big girl of the three, Nikki (Nicole Noel), is the oldest pup at twelve years of age and our first of the clan. Technically she is no longer a puppy. I seem to call all dogs "puppy" because they maintain an innocence which is never lost as they hold onto their lovable qualities as little ones. She's the boss and fairly large in stature for a golden even though she's packed on a few extra pounds. You need to spoil your pets while they're still with you. Besides that, extra care and attention make their day. How fantastic is it to make someone's day? She too follows us around the house and yard as if my wife and I were the Pied Pipers of a devoted kinship.

All our dogs have a great desire to be part of the tribe. Being family oriented, they partake in family barbecues and numerous family house parties. They are always very excited when someone enters our house, evidenced by the old girl who will pee just a little with excitement. It's annoying, but we're now used to it and greet our guests with paper towels in hand. They all want their daily loving and spoiled behavior which continue the bonds we've made. As humans we may or may not recognize that life calls out to us either verbally or silently. But either way none of us would ever willingly choose to give up these connections.

There is an unwavering loyalty by a dog toward their human family members. This goes beyond their domestication by our ancestors. It also isn't something that they have learned. This valued trait is bred into the species. Some may be guard dogs defending their home or property from strangers, and still other dogs are protective of their masters. Golden retrievers love everyone, which is concerning if someone were to break into the house. They would treat them as a long lost friend, happily greeting them. However, I've noticed that a dog's loyalty can be personal, as in one on one with its owner, or with the entire family. It may not always be in a protective manner but in their desire to simply be with us as a companion, always by our side.

They are without a doubt the most forgiving creatures to ever live on earth. Dogs have this inner sight to see you as you truly are. While other people may not forgive your weaknesses, a dog will. And if you reciprocate loving affection they too will respond in the same way until their last breath. A dog will *never* leave you.

Prior to having our three golden retrievers, we had another golden among our family. Her name was Maddie. At that time, we lived on a road called Shady Lane. So when we picked up Maddie as a pup, we named her *Madison at Lanes End* for her formal official documents. She was always referred to by her nickname, Maddie. She was an incredibly loving member of our family. Maddie was particularly gentle with my very young children. However, this seems to be the nature of golden retrievers. Somehow, someway this is an instinctual behavior and disposition of this breed of dog. Our three retrievers now have similar affectionate and good temperament qualities as did Maddie. She will be forever in our hearts and minds for gracing us with her presence.

During this period in my life, I was an aspiring musician playing guitar and piano and constantly writing music. Writing lyrics came so naturally for me that I always carried around a notepad and pen for when inspiration would strike. After the family settled into bed for the night, I'd often get up around 12:30 am or so and go into the music studio in our house to work on my music. It was always quiet in the house and my head at that time was in the right place to be creative. Of course I had to be mindful of the clock to get some sleep before my day job. However, regardless of how late at night, Maddie would get up from her bed, or out of ours, and come with me as I'd play music and work on new songs. As tired as she may have been, Maddie always came and would lay by my feet as I worked. I believe it was more than a companionship issue. I think she truly enjoyed listening to music. She was part of the tribe.

Maddie lived a long happy life of over fifteen years, which is rare for her breed. Her arthritis was agonizing. We couldn't bear to see her suffer much longer. This is when life's decisions are the most painful to endure. We knew we had to let her go, for her sake, not ours.

That day when we were forced to lose her was a day like losing one of your children. I remained with her, nose to nose, at the vet until her last breath. It took all my strength to leave her. But we had to do what was kind and right for Maddie. What we do may often tear at the heart, but our hearts are large enough to hold onto precious memories. We agonized over our own selfishness of keeping her as she progressively struggled with painful arthritis. But there gets to be a point when we all know *it is time.*

Human behavior has a tremendous number of angles to it. We can be kind, cruel, selfish, giving, egotistical, sacrificing, noble, and heroic. However, when it comes to executing difficult decisions, there tends to be the choice for us to select the best option for others. However, we are not the only animals who sacrifice for loved ones. Other species of animals have similar bonds or affections for their kin. People do have this ability to fight against their selfish desires for the welfare and kindness of others. We do what we do because it is the *right* thing to do.

* * *

Spending almost my entire young life along the south Jersey coast in the town of Wildwood, New Jersey, I felt gifted to have this opportunity. I've spent a great deal of time in and along the ocean. I feel that I'm more of a water person than a land animal. My wife and I brought our young kids to the beach before they could walk.

Our second favorite locations after leaving Wildwood when I was a teen were the towns of Cape May and Avalon, New Jersey. Each of these towns is approximately sixty plus miles from our home in South Jersey. The amazing thing was that no one in our family took that short drive from home to the shore for granted. Every trip we made was like the first time ever. Even when my brothers and I were young, we all loved every moment at the shore. There are so many memories of when I was very young living along the shore during the summer. I will never forget them.

As I grew older and eventually married, my own children got to experience the same things. It was sad enough when we needed to sell my grandparents' longtime Wildwood summer shore house. I was in my early twenties at this point in time. However, situations being what they were back then, the selling of that house was done out of necessity rather than by desire.

Although we spent countless day trips to the beach, my family eventually spent a great amount of time at one of my closest friends' parents' house in Avalon, New Jersey. We typically would go down for two weeks out of the summer just as a family. Eventually others would follow to enjoy the summer home. My good friend Ken had a boat slip with his father's boat docked all summer long in the backwaters of Avalon near their summer house.

While visiting Ken and his wife one day, they asked to take my wife, me, and my two children out for a boat ride in the morning. It was early in the day with little activity along the backwater area. As we were motoring out parallel to other boat slips, I saw a dog frantically swimming in the water near its owner's elevated boating dock ramp, apparently designed for the dog to get into and out of the water. There was no one in the area to see the dog in the water.

It was an older golden retriever swimming tiredly around the dock. We watched momentarily, wondering how and why this dog was desperately swimming in circles. There was no way out of the water for him. We were about fifty to seventy five feet away from the dog and I asked Ken to stop the boat.

Surveying the scene, I realized that the dog could not get to safety onto the dock. I dove into the water and swam over to him. By this time the dog was exhausted. Who knows how long he had been in the water. The golden retriever was whimpering, panting with expected dread. He was utterly panicked by fear and frustration about how to save himself.

His eyes gazed at mine, and I could tell he was begging to be saved. I felt his pain and desperation to live. I've been there before. This unknown golden retriever was as important to me as my own family members. I'd do anything to help him.

Fortunately, I saw a wooden ramp that was in an upward raised position on the dock. I was sure the dog probably used it to enter and exit the water. He most likely swam in the water while his owner was present. I quickly shimmied up a slippery, slimy piling to the dock and lowered the "dog ramp" so that he could get out of the water. This ramp had to have been installed to allow him to get in and out of the water when it was lowered. That morning he may have fallen or jumped into the water on his own, not realizing that the ramp was in an upward position. The dog's owners obviously did not realize that this had happened. It was by no fault of the owner. They were completely unaware of the circumstances.

I dove back into the water and needed to push the poor dog up the ramp to safety because he was so exhausted. I'm not sure who was more content, him or myself. His eyes focused on mine. From the dock, his face showed me his gratitude and love for my efforts. I then swam back to our boat waiting for me and climbed aboard.

At that moment the owners came out of their boathouse wondering what had happened. Little did they know that I saved their dog from possibly drowning that day. I was more than pleased that the dog was safe but felt no better about my actions for something that I would naturally do. There was nothing noble about what I had done, nor should anyone be recognized for what they are *meant* to do. These types of small but significant acts of kindness are simply ordinary expectations of what it means to be human.

Humans do so many things every day to save lives. People very often never really know what they say or do to impact others. But it just may save them in ways in which we'll never understand. I never thought twice about attempting to save that animal or ignore it and go on. There was no choice to make. It was pure instinct rather than calculated risk or decision about what should be done. No way did I feel proud of this action, just compelled to do what I needed to do. This type of behavior is a human trait in response to emergencies and disastrous situations to rescue or save others. For most, *it is what we do.*

* * *

Unfortunately, there are many ruthless actions some people undertake toward one another. Understanding the motives of the one and only race of humans will be an ongoing infinite task. It will constantly evolve. For whatever reason, whether it be personal, religious, philosophical, or altruistic, helping others is a noble human behavior. For me, it would be an unacceptable act to not assist another in need, to turn away, or to pretend that someone else will take care of the problem. People are meant to defend one another from potential danger, just as a faithful dog would defend the master it serves and loves. We are here for many reasons and for one main purpose, but I'll let you figure that out.

People seek comfort during difficult times in their lives. Little do we consciously realize how much a dog or other pet may comfort us. Humans don't always need verbal reassurance or discussion to feel content or relieved after a tough day. Simply seeing and feeling a dog's happiness dissolves any past irritation or despair one may have.

When I come home there is no other creature that greets me at the door like our dogs. My wife and I could go out to the store for ten minutes, and when we return they greet us as if they hadn't seen us in years. They are just so happy to be in our company, as we are to be in theirs. All they really want is a hug, a pat on the head, or a belly rub when we come home. To them, it's not home unless we're all here together.

* * *

When my kids were young, I'd get a similar greeting when returning from work. People want to share their daily events or thoughts when we're separated. Children have an exceptional way of expressing their feelings when they've missed you. Don't ever take what they do each day for granted. It only comes along once. Since dogs don't speak, they turn themselves inside out to show their emotional love for you when you come home. When three dogs are greeting you at the same time, it can be difficult to give each the attention that they deserve. However, if one of them thinks they need an extra belly rub, they'll let you know.

Even though we have three golden retrievers, each has their own personality. They are so much like people. Our oldest, Nikki, the female, now has that typical white haired face that older goldens have. The middle child is Indiana, named after my young son since he so much wanted to be Indiana Jones. Indy is technically his dog but has always remained here at home even though my son now lives in a house of his own. The youngest, Dakota, also known as Cody, is a male which we kind of adopted.

We acquired this nine week old puppy from one of my wife's friends. They had just had their first child. Having a young pup and a new baby was too much for them to handle. Knowing how fond we are of dogs, they asked if we would take him. We renamed him Cody. This latest addition to our family has a unique personality. He's full of piss and vinegar half of the time. He likes to sleep under his softly padded quilt, upside down and kind of sideways in his crate. I have pictures to prove how strange he really is. During early spring and fall when he's running around in the yard, he'll often lay on the pool's safety cover, which allows pool water through, so that he can cool off. He then goes tearing through the yard before trying to come into the house filthy, dirty, and very wet.

My wife and I, our kids, and grandkids each have unique relationships with all three dogs just as we do with other people. I have to think about how our early ancestors missed not having the ability to establish links like that with other creatures other than their own kind. It is interesting how we have evolved to bring into our lives the attention and care we give and take from our pets. Right now I'm not sure if we have our dogs or they have *us.* Either way, it works, and I wouldn't want it any differently.

It is fascinating how dogs and other animals have become part of "*the human experience.*" Our behaviors have certainly changed from taking care of them simply as pets to having them as part of the family. When did human behavior change for this practice to become part of our nature? More importantly and interestingly is *why*? We rarely recognize shifts in evolutionary change. We always think that it is going to be physical rather than behavioral. Apparently it is both.

The eternal affection of a dog among human life goes beyond words. There are too many personal descriptions that we could incorporate into why they have become part of our lives. As much as people need to care for other living creatures, the more we see that life calls to life. The noblest act is still for personal satisfaction, which is not meant in a selfish way. It is most likely an innate behavior for the better good of the recipient being ourselves. There is nothing wrong with finding our own pleasures in life. Still, a person with a moral code of conduct will do for others even at the sacrifice of oneself. Dogs will always be loyal toward a well deserved owner. They are a perfect example of how we can care for these loving creatures and experience their unconditional love in return. This is what we seek from our own kin.

The one thing that most humans fully enjoy doing is laughing and finding humor. Our pets often have ways to entertain us, and dogs certainly provide more humor than most adults. It would be great if dogs could talk because I'd like to know what's going on in their minds. We take human speech for granted. Humans think we use language to understand our thoughts. What would a dog use to understand its own thoughts? They have a language of their own, but what is happening inside their minds?

When I look at how my three dogs act, I assume they're thinking using words as I would. However, I'm totally wrong. Nonetheless, dogs do understand some human words which they learn and recall over time.

Here is a case in point. Cody got the scent of an animal in the backyard one day. I'm watching him rush to a bush near the backyard privacy fence. He's racing around in and out of the bush, sniffing and smelling the surrounding ground. I figured that he caught the scent of a bunny. If Cody could talk, he'd probably be saying, "I smell bunny, I smell bunny."

I see the rabbit sneak out from the bush and up the hill to escape my unknowingly oblivious dog. Meanwhile my genius dog is still racing around the shrub. So he thinks, "OK up the hill." Therefore he runs up the hill. "No down to the bush, no up the hill, nope, back down again." He never caught the rabbit. The very last words in his mind were probably "Oh, $%#@!! I'll get that bunny next time."

For the next 15 days Cody would run back to that same bush, sniffing around, thinking the bunny was still there. By then the rabbit was on the opposite side of the privacy fence giggling at my brilliant golden retriever bunny hunter.

As ridiculous as dogs sometimes appear, they're pretty clever. As I discovered one afternoon, I mistakenly didn't completely close the kitchen pantry door. Several minutes later, when I entered the kitchen again, I noticed the door completely open. Sure enough, the half full box of large dog bone treats was nearly empty. I looked around and immediately saw Cody, also known as "*Damn it Cody*," so named for a reason.

He had an expression on his somewhat guilty face. You can always tell. They're like kids who don't speak. I knew it was him. Wide eyed, ears slightly back, not looking at me directly in the eyes, trying to go about his business as if nothing had happened. I swear I saw him later sitting in my reclining leather chair rubbing his big belly full of bones which he had eaten. Needless to say, one hour later it was 5 pm and he's giving me that stare for supper.

There is a ritual in our house about when it's time for the dogs to eat their daily meals. They have breakfast and dinner. We try to stick with the same times of the day to feed them.

Dinner, or supper as I prefer, is at 5 pm. They all start eyeballing me about ten minutes *before* 5 pm. The dogs know when it's time to eat because I get that unwavering stare from all three of them at 4:55 pm.

Can I please have 5 more minutes to finish the early news program before I have to get up and feed you? Of course, their hypnotic stare at me to get off my ass and feed them immediately at 5 pm became a daily routine. If my wife and I come home with take out dinner, we always feed them before eating ourselves, no matter how hungry we are. With their sense of smell, that would only be cruel if we ate first.

Approximately fifteen minutes after eating their supper I would get the "poop stare," and they'd want to go out back and do their business. It was much easier raising kids. The one thing that my wife and I don't allow is giving the dogs "people food." However, that does not stop them from giving you the "stare." They always eat before us, no matter what the circumstances. However, you wouldn't think so by their actions. Still, they are expecting to get some of our dinner or snag an accidental morsel that may fall to the floor.

I know that dogs have at least one hundred times the sense of smell compared to humans. The ability of smell is a powerful motivator for many creatures, including us.

When Nikki was a young pup, she and I would have an evening ritual of racing upstairs to jump into bed to get into *my* spot for the night. She would beat me up the steps and always lay in my area on the left side near my pillow. I'd come in and ask her to move. I would have to physically attempt to relocate her to the rear of the bed. She would complain in grunting, low growling, but in a non threatening manner about repositioning her spot. This happened every single night. If I didn't know any better, I'd swear it was a game she was playing.

If you think about this, it is similar to what parents and kids may do when it comes to testing your resolve. The key is to be consistent. Nikki would eventually come up to where I was laying for a final belly rub before going back down to the foot of the bed. Children will push their independence on their parents as they mature, and you *want* to see this happen. It is a developmental process for self sufficiency and independence, for which we continually guide them until early adulthood. It is a need and gift for us to fulfill this mission. I will say that we're usually more of a pushover when it comes to giving in to our pets than our kids.

All our dogs love to go for a ride in the car. I can't imagine what they're thinking, but they make every attempt to get into the car. It's a good thing they don't understand the word *neuter* or two of them would never want to get in. We purposely don't use the word Vet around them, although every time they would go willingly.

Each of the three acts differently in the car. "Damn it Cody" is nonstop movement in the back seat. Nikki finds a place to lie down, and Indiana calmly sits by the window peering out at the scenery as it goes by. He looks like a little man sitting in the back seat. One of these days I'm going to place a pair of goggles over his eyes and a baseball cap on his head and drive around town in the convertible with the top down. They all enjoy either sticking their nose out the window or hanging their heads out for some time. It's a good thing humans do not rely on their sense of smell as dogs do, or your passengers would all have their heads out the window too. Cleaning the dog drool off the side of the car as they do is no easy task. I pity the poor driver behind me.

I would love to know what goes on in a dog's mind. There are basic innate needs and desires, but there is more going on than that. They are very unlike humans in complex motivations and ambitions. A dog will evidently show you what it wants and how it feels. I'm not sure if most people recognize animal behaviors as well as they recognize human behavior, because dogs do not verbalize like people. Because of this, they are often misunderstood.

Imagine attempting to communicate with someone from a different country who has a different language. The basic questions and responses are most needed. The same can be true with a dog. Although they may be able to read much of our body behavior, they don't comprehend language as another person does. And we don't always read them clearly. What is the dog asking for when it approaches you? After a period of time, we will both understand body signs along with dogs comprehending basic words or hand signals.

My wife and I always utilized hand signals with verbal commands when training our dogs. The two methods seem to work well, and the dog learns faster. Trust me, it can help eliminate a great deal of frustration for each.

Tell tale signs in their posture, eye contact, and body movements elude many people as to what a dog wants or needs. Indiana will constantly be touching my foot with his paw as I sit in my family room chair. If I move my foot, he instinctively reaches for it until he finds my foot. He'll gently lay his paw on my foot once again. *Contact! Connection! Love!*

Cody, not realizing how large he is, insists on jumping up and sitting on my lap while I'm in my reclining chair as I watch television. He's so large that I can barely see the 55 inch TV ten feet from me. He buries his head in my chest as a hug and likes me to rub his face and neck. Nikki leans next to my chair so that I'll rub her soft head. They want touch and affection from us just as people want this from each other. They have so many related human behaviors, and one is to belong with their now adopted tribe.

Humans treat their dogs and other pets well because we choose to do so. I wonder if it is a little more than a choice. We may be compelled to love and comfort them because we can, and over the years it became inbred in us. The fascinating question is always why. Human nature has changed and will continue to change as we evolve. Let us hope that we don't lose those special qualities that make us totally unique among all other animals.

Dogs have a way of helping us keep our feet on the ground. If one goes astray, bring in a dog to remind us how compassionate we are meant to be. Service dogs have helped so many people over the years. Our three golden retrievers can bring me back to recognize how good things really are, and suddenly life doesn't seem so difficult or troubled. It's OK if we momentarily lose our way because we are only human. Sometimes it just takes a dog to remind us.

Dogs are smart animals. They have higher intellectual abilities than wild animals such as their close cousin, the wolf. Being around humans for at least seven to eight thousand plus years has given them an advantage over wolves. There is quite a vocabulary of words that our dogs know very well. They all get so excited when I say, *"let's go clean your yard,"* meaning I'm going to pick up the poop. Oh yes, that's my chore, not my wife's.

You'd think that they were all going on an adventurous journey by the way they dance around to go out back with me. Sometimes all I need to do is go to the foyer and pull out a specific green plaid thermal shirt from the closet, and they know we're going out back. They all do their happy dance as they tug and pull at me while I can barely get my thermal shirt on. It's a big deal to have no poop in the yard, which by the way is almost a daily event with three big dogs. Big dogs mean *big* poop.

Retrievers have a very good understanding of human vocabulary. There are certain words which we need to be careful of saying in front of the dogs such as ride, walk, snow, ball, and snack, just to name a few. Their knowledge of human language sometimes appears to exceed most four year old kids. They can also associate the meaning of words like humans. If I say *"snack,"* they know it involves food. Or I could say, *"Who wants a bone?"* and they still get all excited. Regardless of the food source, their complete intention is to fill their bellies.

To watch their diets, my wife and I substitute traditional treats with healthier carrot sticks. You'd think they were getting a steak dinner if you ask, *"who wants a carrot?"* During the hot summer, an occasional ice cube satisfies as a treat. It is very difficult to take all three goldens on a walk at the same time without having a handful of treats to keep them on course. One scent of a bunny, and they will be going in all different directions.

Several years ago, a local news channel was on TV during the winter. We were all lying around watching the weather report. My dogs were half napping. Suddenly, the weather reporter said the magic word. All of the dogs from their groggy nap heard the word *"snow"* from the weatherman. They all got excited, their heads and ears perked up, and they began to look outside the glass sliding doors to the backyard. This was a word that they heard a news broadcaster say on television. They are all very familiar with the word because we use it all the time before we go out and play in it.

Every snowfall in winter is a fun filled day of play for the dogs. Indiana and all the dogs love going out in the snow. However, Cody in particular loves to roll onto his side and kick his legs in the air, then roll over again on the other side to do the same thing. He'll then shake himself off and roll around again in the snow.

We have a small hill in the backyard. When our kids were young children, and now our grandchildren, we would spend half the day sledding down this tiny hill. It is a very small hill but steep enough for a little fun. Indiana also liked to *"sled"* down the hill. He would sit down, pick up his rear legs, and slide down the hill on his rear. I'm OK with that, as long as he does it outside in the snow. We don't want him practicing that trick in the house.

Snow was one of the best times to have fun with our dogs. Neither one of them wanted to come in, but we eventually insisted. Each of their legs would be covered in *"snowballs,"* as the snow would stick to their fur. It was like taking care of kids who played outside all day during the snowfall. Cleaning and drying the fur of each of the three dogs was a two person job. But we didn't mind. They were our four legged kids.

The summers are no different for dog fun out in the backyard. We have a fifty foot long in ground pool, which our golden retrievers believed belonged to them. Indiana and Cody were usually the first to jump in as soon as we took the safety cover off in spring. The old girl Nikki just liked to get her feet wet. Indy would sit on the steps halfway in the pool after a quick swim, and *"Damn it Cody"* would not stay out of the pool at all.

The very first thing in the morning after a quick pee he'd jump in for a swim. So by 7:30 am he's already wet for the day. Do you know how many dog towels we'd go through in a day? They were the old torn ones which we keep for them. When people come over and see these old wet dirty towels hanging on the towel pole, we quickly remind them that they are for dogs, not people. If we didn't say anything I don't know what people would be thinking about us. But there is no question that the dogs know how to have fun. It's in their nature just as fun is built into the human genome.

In today's world, "fun" has taken on a whole new meaning than it did years ago. Our poor ancestors, who struggled to survive, knew nothing of enjoyment and pleasures outside the daily tasks of attempting to stay alive. Human behavior involves a limitless range of interactions among our species, including bringing our pets into those activities.

Who would have ever considered creating the popular "dog park" years ago as it is common today? We include our dogs in many more aspects of our daily lives. Many of us look forward to coming home from work and taking a long, peaceful walk with our dogs. My wife or I just need to ask, *"who wants to go for a walk,"* and all three go crazy for a run in the park with us.

There is an elementary school just through a wooded path from our house to the school. We would occasionally take the dogs for a run there after school hours. It was quite large and fenced in, which didn't really matter because our dogs would come to us when called. Unfortunately, we needed to end that routine because of the numerous Canadian geese that would frequent the area and leave enormous amounts of goose poop everywhere. Initially our dogs thought they were in a field of dog brownies to be gobbled up. We very quickly ended that.

The enjoyment for them is not only getting out, but physically being with us in whatever we do. Dog parks supply endless smells for them to sniff. There is nothing more stimulating for a dog than to sniff new smells, especially when the more horrible the smell, the better. It stimulates their brains and gives excitement with newly acquired environments. Not only do they have a vast amount of space to run in open fields, but also the opportunity to meet new dog friends. And we all know how dogs greet and meet new dogs. Thank God humans did *not* pick up this trait.

So, if you're reading this wondering how this chapter regarding the eternal affection of a dog fits into the subject of a discussion about human nature and human behavior, the answer is simple. They're just like us in many ways when it comes to basic drives and desires. They are attention seeking, do anything to keep their bellies full of food, and wish to love and be loved by others. Basic. Everything else is extra.

Whichever ancestor was truly responsible for domesticating the dog seven thousand plus years ago needs to be commended with unending gratitude for the relationships that we have today. I could not imagine life without a dog in our lives. Somehow, I can't imagine them being without us. It was destined to be, for which I do not want to know the reason why.

Sometimes the presence of a dog by your side can be more comforting than words spoken by any person. Their quiet emotional affection touches our soul through subtle actions that tell us, in a unique way, that we mean everything to them unlike any other creature. Dogs can read the emotional state of people better than we can read one another. And it all happens without a spoken word.

The softness of their eyes as they gaze at ours, the gentle wag of their tail for no apparent reason, and the touch of a soft paw for physical contact become a connection recognized only by the heart. I believe we are eternally *blessed* for this.

A dog may also be the only animal that can feel the reciprocated love from us through our gentle touch and desire to be with them. Never have two different species coexisted in such a harmonious dedication to each other as a human and a dog. In my many years of living, never have I recognized this so profoundly.

For a child, a boy and his dog are like brothers to one another. The loss of either leaves a lifetime of emptiness which can never be completely refilled. A dog in many families is not simply a pet. They are part of the family, with an identity that makes them unique and an intricate part of how the family collectively functions. Dogs want two things: to be loved by people and, in return, to demonstrate love to us. Oh, and to chase bunnies. Therefore, who could dispute the eternal affection of a dog?

Chapter 8: The Telling of Tales

The telling of tales from generation to generation is done to keep our human history alive and, in the foreground, to remember. It is important to know where we come from and how we may wish to proceed toward our future. These tales are more than just stories to occupy young minds. Most people believe a tale told is nothing more than a fictitious story. I like to redefine the meaning of a "*tale*" as something that is a piece of truth, a story, an embellishment, and a degree of factual meaning to be told and passed on to others. It is a verbal description of events that pertain to the personal or global history of our species. The impact of these tales molds and reshapes thinking and concepts about how we handle new and forever changing events. It also reminds us of how we've changed or what we should hold onto. In any event, most people are only interested in where they are going or what they want out of life. However, not knowing about some of the documented events of our history can lead us onto disastrous paths previously traveled. Besides, the telling of tales can be most hilarious when retold.

The telling of tales has not only historic value but also the satisfaction of one's own personal history, which should make us feel proud. If you're looking for an absolute truth, you probably won't find it in tales, stories, or even sometimes in written history. All of them lend to bending of perceived realities. So, *tell* your tales, *tell* your stories, and *tell* your history. It all leads to the potential flaws found in human nature. To know and understand life, you'll need more than tales. You'll also need to live it. At the very least, we may be able to find a sliver of humor about all of it. In the end, people choose whatever truth makes them content.

Human behavior has two choices: one, dwell on the past with resentful regret, or two, look to the future with a hint of optimism that will benefit you or our species. The choice is ours in how we move onward. We could complain about how dreadful life can be *all* day long or get off our ass and make a difference for ourselves and others. People who get on with it rarely forget about the past. They use it to their advantage. Early ancestors would never have survived without remembering what was passed from generation to generation. They then went out and used it. We use these memories for two purposes: first, to learn from past experiences or knowledge gained, and secondly, to enjoy the memories of the people and events we've encountered. Humans can only do this by *the telling of tales*. If you prefer, then it is the telling of the perceived truths and events deemed worthy of keeping and passing on those memories to the next generations. I believe that the "tales" enlighten us to the joy that overrides our struggle to make it one more day.

Human nature has the need to keep the past alive. It was a survival technique used by our ancestors. Somehow, they found the wisdom to share valuable knowledge with their offspring. It indicated an expanded range of understanding to think ahead to the future for their kin. Insight into how to survive was a necessity to pass along to new generations via verbal communication or demonstrated skills for them to be successful. The interesting subject to me is why our ancestors wanted their offspring to continue in the first place? Why should they really have cared? Perhaps it was because it was no longer a personal need to survive but rather an urgency for the *tribe* to carry on. There seems to be an innate drive that developed for the passing of knowledge or telling of tales that had evolved in our species. Many wild animals have acquired behaviors of teaching or protecting their young through their own evolution and need for survival. This behavior is genetically ingrained to care for their own kind, live as a herd, in a pack, or as a group. Knowledge or information can be shared in many different ways, but usually it is for a significant purpose.

Although we must remember that human behavior is all about the telling of our past or elaborating on current events. Everyone loves to hear a good story, even if it isn't entirely true. Gossip and false statements lead to further tales which may enhance the story and occasionally poke innocent fun at the recipients. The wilder the story, the more interested people seem to be. Hence, conspiracy theories were created and thrived upon among many misguided or naive individuals. Based on some of the things I've heard, life isn't that mystical. Funny? Humorous? Yes! Perhaps some stories were intended in jest, to spark our wild imaginations and insatiable curiosities. But let's not make fun of what others perceive as possible truths. Who are we to judge something perceived to be true? Keep your opinions to yourself. Some things are not worth the debate.

Of course, the telling of tales in no way should humiliate or degrade someone's reputation, whether it is true or untrue.

There is a line not to be crossed when it comes to sharing information or stories. Teasing family members about some of their idiosyncrasies is a given in our household. Funny things that happen within the family stay in the family; it's like whatever happens in Vegas. However, gossip does not fall into any category of having value in the telling of tales as I see it. What we actually pass along as value must be essential or relevant to our kin. This has been demonstrated by our history for thousands of years. Sometimes the telling of tales becomes longer and longer. Who hasn't explained a "fish catching" story? How *big* was it?

There are many ways in which we can pass along information. Verbal communication, whether it be from story or through detailed facts, is important for learning or understanding our past and making realistic decisions about our future. The oldest and most ancient way of communicating ideas and knowledge is through demonstration, not necessarily verbal sharing of information. Very early human ancestry did not have our current expanded range for verbal expression. From an anatomical point, we would never have been capable of speech if the hyoid bone in the throat did not lower to produce the sounds as we can today. Spoken language would not have been able to develop. Evolution played a significant role in that physical change. Therefore, time for physical and mental change enabled our ability to advance.

I suspect that at some point in time, our ancestors many years ago told one another stories to elaborate a truth. That seems to be human nature. So, the telling of a tale is not just fiction but has its base built upon truth as one perceives it. I could be wrong, but I don't believe Neanderthals had a sense of humor when it came to communicating with one another. Yes, they did have very basic verbal communication. That being said, this might make our own history just a little more humorous than it already is. I can hear it now: one Neanderthal saying to another, "Hah, hah, my spear is longer than your spear," originating the saying that "size matters." Or perhaps our Neanderthal cousins saying, "Hi honey, I'm home. Oh... sorry, Floyd, I thought you were my wife." I know, with all that facial hair and everything. Not so funny. The truth is they were rather beastly in appearance by no fault of their own. I highly doubt that humor was a priority to them. I'm sure their daily routine would have been one of dread to find food, stay warm, and keep all members alive for the hunt and continuation of the species.

I doubt very much that the non-politically correct statements of today would have started with those ancestors. If so, then it is a shame that we did not evolve past the insulting, humiliating humor that often dominates much of our daily conversations today. I love funny stories, but a line needs to be drawn when they become totally disrespectful or inappropriate toward others, especially when it comes to one's physical appearance. Behavior is the exception. Today, if we act like assholes, it's probably because we are, and should know better. So, stop it! Remember, we're civilized human beings. Or let's hope we're getting there. At least pretend you are.

Telling stories has more of an effect for entertainment purposes or for relaying family history around social gatherings. That is the true value of the telling of tales. I must say that storytelling sometimes places certain labels on individuals. This is by no means meant as an insult. Often, it is done in good fun. And what would life be without good fun? We have all been subjected to practical jokes and humor that pokes fun at ourselves and others. It is the inside family humor that makes for enjoyable tales and fond memories. Why don't we keep that measure of "fun" between family? If everyone else's family is like mine, then it is all in good humor—no insults.

*　　　*　　　*

One family tale that is still repeated today is about a trip to Hershey Park in Pennsylvania. Almost every year, my wife and I would take our kids to this great amusement and entertainment park. This was usually a two-day excursion. There were roller coasters, small rides for kids, many places to eat, and entertainment shows every hour or so. Kids and adults enjoyed the park. We looked forward to times when they were planned as a family trip.

One year, while there, our family went to see one of several different entertainment shows in the afternoon. It was an outside theater. We were sitting in the front row. My son, who was around six years of age, was seated next to me, followed by my wife and daughter, age nine. During the performance, these adult entertainers came out appearing to be *riding* large flamingos. Now, you had to think about this. The people's actual legs were the flamingos, and the legs over the saddle were not real. My son and wife looked at this with absolute astonishment! They couldn't figure it out. My daughter and I knew right away. She and I looked at the opened mouths and surprised, confused expressions on my wife's and son's faces. My daughter and I looked at each other and started crying with laughter at their bewildering looks. My wife and son had the most bewildered facial expressions as if they had seen *something* miraculous. Their minds just couldn't compute the mystery in front of their eyes. It took several minutes before they realized what was going on. In the meantime, my daughter and I had the greatest hysterical laugh of our lives. It will never be forgotten.

It is a tale that has been told for the last thirty years among the family. It changed our lives! Ever since that day, my wife now has an affection for these unique birds. She will routinely buy plastic flamingos to decorate the yard and garden. There are flamingo pool floats, flamingo pillows in the Florida room, and even a flamingo umbrella for the beach. We still laugh about that story. It comes up every time we see something flamingo. This is the type of tale that most likely will be forever told within our family. And I am so pleased that we have this small piece of family history and camaraderie between the four of us.

* * *

Many years ago, a man was talking with a friend who was a coworker about how each occupied their free time away from the job. The work they each did was demanding, exhausting, high in volume production, and technically challenging. Everyone who worked there left at the end of the day and would find different and sometimes unique things to reboot their lives prior to getting up the next day to do the same kind of work all over again. People's own individual time should always be self-appreciated and utilized wisely. Many of us don't always have enough time to do the small important personal things in life. They should be noticed and we should be grateful for them.

After work during summer, the gentleman would go home and swim a thousand yards or so in his pool and then lay on a pool float in the late afternoon, gazing skyward at the clouds, almost as a daily routine. The peaceful view of watching the formation and movement of clouds, including their changing shape, coloration, and observing the "*death of a cloud*," was always recognized. The death of a cloud is when it slowly dissipates into thin air and is gone from view. Eventually, that cloud will fade into a clear open blue sky. As time continued, another cloud will once again do the same thing. It can be easily explained but appears as a phenomenon.

He told his friend about his skyward gazing and recognizing clouds while lying in the pool, and it drew an instant response from his coworker. He said, "*You're a Cloud Watcher*." And the inside joke about the fascination of watching something relatively meaningless remained somewhat humorous between the two of them. However, a cloud watcher has special meaning; recognizing the small irrelevant things in life that actually have value and making connections between that which appears inconsequential, to the relevance and significance of daily events which are meaningful. Please watch the death of a cloud, as it slowly fades away and becomes invisible to the naked eye. It is gone, destined to become part of something different and bigger in that weather system. Unfortunately, this type of thing happens to some people with their lives. They fade away without ever being recognized or truly impacting others. Sometimes it is by their own choice. These people are like clouds that exist for a while then fade away without anyone ever truly noticing that they're gone. How sad is it for this to happen to anyone? Fortunately, clouds do not have feelings, but they are part of a complex cycle in the flow of precious water. I'd like to think that every person has a part to play in the act of being human just as clouds have their role.

The cloud watcher had spent a lifetime finding out who he truly was and continued to spend his life attempting to accomplish it. He recognized that he still had so much to do, both personally and professionally. As busy as he may have been, he always took time to look up and gaze at morning and evening clouds. It is a simple reminder of the small things which may have the greatest personal value. Great things do not need to come in large packages. It could be the changing of the wind of an approaching storm or a sunset never to be forgotten as you sit with a loved one.

The cloud watcher wasn't created; he was born that way. It sometimes may appear that a human has an undisclosed pre-planned potential agenda based upon their abilities or capabilities. It may simply take the initiative to apply it. This may initially seem idealistic or optimistic, but we've seen so many possibilities and accomplishments in almost everyone. The world is too complex for numerous successes to occur by *pure* accident. This wouldn't have allowed us to play our role in life. Destiny is not pre-determined. We do have free will, and the ability to choose which path to take. Nonetheless, there are no guarantees. That "*why*" question can haunt some people if they dwell on the results of the outcomes that they weren't expecting.

The cloud watcher sees minute details scattered throughout different aspects of life that can be expressed in a manner for others to see or ignore if they choose. We don't need to show dominance or submissiveness among ourselves as animals often do. Our advantage may be just to become aware of the numerous small, occasionally overlooked details to enhance our ability to become successful. It is difficult to understand the world around us; therefore, an awareness of the little things may help create a fuller, detailed picture. For one piece of a puzzle could never show you an entire image.

The cloud watcher came to be in his early teens. As a musician with guitar in hand, writing music and lyrics, he expressed himself as he viewed life. There was an awareness that enlightened and enhanced his creativity. The gift of enlightenment can answer basic riddles. From a wider viewpoint, questions on the meaning of life arise. It makes one believe that we are a small speck of life in a vast universe. The questions are sometimes more intriguing than any answers. This is the development of philosophy. The age of the cloud watcher began to emerge.

The cloud watcher knew his strengths and was more than aware of his weaknesses. He denied nothing about his character. However, being of stubborn English and German descent, he refused to accept circumstances dictated to him without placing his own will on attempting to change or affect potential outcomes. That was his nature. Of course, in addition, his Irish heritage also made him a lovable character, especially at parties. This type of jovial attitude has always found ways to make good times out of the harshest moments. Nonetheless, we are the product of our parents, and this can make us out to be our very best. The cloud watcher knew where to give credit when credit was due.

An extremely important part of the cloud watcher's desire was to remain close to family during frequent gatherings to listen to the telling of tales by the elders. They were precious treasures never to be forgotten and to pass on to future generations. The tightly knit family bonds and communication laid out the groundwork for the cloud watcher to be expressive in his writing of music. Lyrics flowed out of him like water from a tap. It was a way for the cloud watcher to tell the stories in his mind and how he felt. Expression and the need to be heard is a human quality. For those who understand, it goes beyond simple recognition. As open as people can be, there is always a corner in our hearts and minds that belong just to that individual. It is not a secret; it is simply something personal which must remain solely with that person. It most likely connects with the desire to still remain an individual. One must remember, whether they like the idea or not, we are only human.

When the woman of his life came along, and they were married, it was not by accident but by fate. It was meant to be. That, he was sure. The offspring which followed further enhanced his worthiness as a human and his ability to truly participate in life. He could not be who he was without them. This made him eternally grateful. The cloud watcher wonders if this is how it is for all people or if he is just so fortunate. As good as it was, there were also the deepest, darkest places of the unknown not yet traveled. But that is because there cannot be light without darkness. Often, the light must be *earned.*

There is often a time in one's life when they will lose their way. It doesn't matter about the success or potential failure we experience. It is a point which occurs when we seem to be in shadow. Whether this is a product of inability, focus, or purpose, we feel lost on the path traveled. Often, we look for pictured images while ignoring details. This is the time when we need to step backward momentarily and refocus. Take the time to recognize the things that are really meaningful and important. Your life could be on the line. The saddest thing might be losing yourself. The cloud watcher made time, whether it was convenient or not.

There is an old saying about stopping to smell the roses. In a way, look past your busy schedule and watch sunsets. Hear the birds singing in the morning. Watch the moon travel across the starry night while sitting by the outdoor fire pit. Gaze at the clouds. It will reboot your conscious mind to focus on that which we must do and remember the importance of our existence. Human nature doesn't come with a set of rules to follow. There are no instructions. And there is no specific course to guide human behavior. That is why living life can be so varied and often so very hard. It would be great if we were to just follow a set of guidelines, but that would *not* make us human.

The cloud watcher has many tales to tell. He has established himself as to who he is and knows what he must become. He has lived many lives with multiple lifetimes of experience. With the knowledge gained, he *knows* the direction that he must go to fulfill his obligation to this life. As a cloud watcher, he *knows* what must be passed on to his kin. Someday, as he will fade away as clouds currently do, there will be no regrets before his passing. He has learned and continues to understand what life expects of us. Besides, we never really go away. And that is a most comforting thought. If you haven't figured it out by now, I am the cloud watcher.

Chapter 9: We Are Only Human

The Out-of-Africa theory of human evolution states that *Homo sapiens* left Africa approximately 60,000-plus years ago after being there for some 200,000-plus years and migrated north into Europe, the Middle East, and eventually eastern Asia. From there, humans traveled south through Malaysia and then into what is now Australia. Eventually, humans took a northern route across the Siberian land bridge into North America. Two theories exist on how we populated North America, but we finally made our way south into Central and South America. In the Out-of-Africa theory, *H. sapiens* entered Europe where Neanderthals had lived for thousands of years. As our human ancestors "*invaded*" Europe, the theory claims that we just outcompeted our Neanderthal cousins. Eventually, the harsh ice age conditions, along with competition for food and space, forced Neanderthals into extinction approximately 28,000 to 30,000-plus years ago. *H. sapiens* remained and dominated the region, being highly skilled with their weapons to hunt, having advanced communication abilities, and employing sophisticated measures to survive.

A second theory surfaced after a scientific discovery that suggests another method of the possible demise of Neanderthals. They weren't eliminated by *H. sapiens* but had occasional close interaction with modern humans. It is suggested that, based upon Neanderthal primitive spears for hunting, a complex language method was used to coordinate the hunt for food. Strategies were needed, and this required intelligent communication and language skills. These genetic cousins of modern humans were not dumb brutes who reacted strictly on impulse as many had once believed. This second theory suggests that modern humans intermingled and coexisted with the Neanderthals to some degree. There is evidence which clearly indicates that these two similar species intertwined to produce offspring. I feel certain that some violent and deadly conflicts would have occurred between the two groups from time to time. However, there may not have been as much contention between the two early humans as once considered. After all, violence and killing seem to be embedded into our previous and current nature.

Neanderthals inhabited many remote corners of Europe, from the central regions to the far reaches of western Portugal. It is here that a fantastic discovery was made in the late 1980s. Scientific evidence indicates that modern humans and Neanderthals intermingled, which included some degree of interbreeding. Approximately three to four thousand years after Neanderthals were believed to have become extinct, the fossil remains of a young child, meticulously wrapped in animal skin and buried with respectful honor, were discovered in Portugal. The bone structure showed a mix of Neanderthal and modern human traits. Chemical evidence comparing modern human genetic code with Neanderthal DNA showed relationships which included interbreeding of the two groups. A hybrid was created between modern human and Neanderthal genes.

Obtaining "clean DNA" from twenty-thousand-year-old remains is difficult at best. One must sort through contaminated DNA from insects and other animals to obtain viable DNA to run a scan and sequence a genetic code. This is no easy task. However, this cast new light on the relationships between modern humans and our close cousins: the *Neanderthals*. There are certainly other theories and opinions about the relationships of our ancestors and the demise of Neanderthals. This will be debated for years to come. The best thing: keep an open mind. Theories are intended to be challenged and revised with new compelling data. Whatever the results, it will not degrade us *all*-powerful, *all*-knowing humans. I love sarcasm! It keeps us grounded. In addition, all *true* scientists would love to be proven wrong. This encourages us to continue working toward a sound truth.

I have learned that all scientific theories need to be viewed as a mixture of several possible events or outcomes. How hubristic is science to claim a definitive and absolute fact about human history based upon theories which currently exist? Theories can and do change with additional information gathered, then confirmed to be correct—for the moment. It is true that theories are not etched in stone; however, science can display some pompous attitude about their acclaimed knowledge, which is occasionally *stumbled* upon by accident. I am sure that it is a combination of truths which eventually settles into reality. Hard evidence can only give further strength to theories. This is only if we are willing to accept potential facts or data, and possibilities which we may or may not want to accept.

We have a good idea about the numerous aspects of current human nature. The fact is, we are destructive, domineering, possessive, controlling, yet sympathetic, caring, nurturing, and altruistic. The descriptive list of our attributes–good and poor–goes on and on. Surely, we are the most *complex* creatures to have ever existed. One could possibly invest in multiple generations of time analyzing and writing about human behavior. Is this because human nature evolves faster than we realize? There are countless cultural differences between us. Obviously, culture is a large part of our nature. But the phrase "*nature*" includes even more than only customs. It is how *we* think, how *we* respond, how *we* understand concepts which are part of human. Therefore, human nature varies among us and is always changing. Certain aspects remain the same while others shift and change with the wind. Even today, we are trying to get the full picture and attempting to understand who we are and how we came to be. But we all know that an equally important question is, where are we going? One caveat that may interfere with our desired goals and search for truth is the fact that *we are only human.* Because of this, mistakes and misconceptions are inevitable.

One thing is for sure: we are definitely human because we certainly say some stupid shit. This is especially evident when we are not trying to. One of our greatest accomplishments is the *excuse.* Actually, our ability for language is genetic. The way humans verbally communicate has saved our species and created relentless questions toward husbands who have mistakenly come home a little late from an evening at the local pub with their friends. The wife can be exceptional when asking for answers to questions that you can't remember. The excuse that there was no clock at the pub does not fly very well. And if you were supposed to bring home milk, forget it, don't even attempt to say the store had run out. Fortunately, joking about how we each behave keeps the relationship going. Tense at times, but all good otherwise. I couldn't be who I am without her.

Humans communicate in different ways under different circumstances. We inform and educate. We joke and poke fun with others, including ourselves. We criticize, are definitely cynical and opinionated, and we express compassion and love, and express understanding and forgiveness. Human behavior is driven by emotions. This gifted ability often screws us. Sometimes we are exceptional at verbally communicating how we feel, and other times it comes out all wrong. It is a difficult task to coordinate the brain and its logic with emotional expression in terms of communication. Bottom line, we don't always mean what we say, and what we wish to verbally convey can go terribly wrong. This seems to go along with a missed opportunity to do right for others. Miscommunication is as frequent as sunrises.

*　　　*　　　*

A main factor about what it is to be human is our patience, or perhaps, lack thereof. I feel confident that this characteristic goes way back through our ancestry. It is often during the most difficult of times that we fall into that failure of miscommunication between one another or misunderstanding of one's true intentions. Consequently, frustration results in people. During a rare hurricane in late October, a tremendous storm scenario included howling winds, rain whipping through trees, and flooding in low-lying areas. This typically creates an electrical blackout. In today's world, nothing is more frustrating than not having power. We need power for almost everything. Even moderate storms in our five- to six-block area of the neighborhood will shut off the electricity. This power outage will occur for most of the day, making it difficult to do almost anything. Electricity would come on for five minutes, then go off again. It was an hourly routine. One in which we did *not* adapt very well. Many humans can be easily frustrated, including myself at times. Patience is worth gold!

During one such storm, my wife and I spent the whole day together watching the news and weather when the power was actually on. Of course, we both observed the declining weather conditions and could do nothing except stare at each other when the electricity eventually went out. When desperation reaches this level, one will notice whatever little insignificant irritating nuisance about our usually compatible personalities. And then, it begins to get picked apart. Yes, we were slightly bored. *Very bored*! This is what people do when there's nothing better to occupy their time. Minor skirmishes would normally be ignored with little or nothing to retaliate or comment. But confinement in a house is like having too many goldfish in a ten-gallon fish tank. Eventually, they turn on each other. The entire day had been irritating due to the power and weather. So naturally, people are not behaving at their best. Each of us was saying to the other, "Don't keep the refrigerator door open too long." Gotta keep that food nice and chilled. And more importantly, you can't let the beer get warm. So, what kind of meal could one possibly prepare without power? We couldn't even have ordered take-out. We did our best to come up with ideas for food, but I wasn't the one who had to go to work that evening. "I promise you a steak dinner when the time is right." It was beginning to get late as my wife got out of the shower in preparation to go into work from 7 p.m. to 7 a.m. at the hospital, two miles down the road. Because of this semi-hurricane condition we were experiencing, I said that I would drive her to work and drop her off at the hospital front door so she wouldn't need to walk far in the storm.

My wife is now out of the shower and hurrying to get ready. This was being attempted when the lights came on for a brief moment. She was just about to dry her hair when the lights and all electricity in our home went out once again. I knew she was irritated even while I was downstairs. I was preparing myself to walk on eggshells as I knew she wasn't pleased to go to work in the first place. They say that the best of us comes out when things are at their worst. *Not always!*

The power is out once again, and my wife comes downstairs while attempting to get as best prepared as possible with no electricity to dry her hair. I thought, do I say something to ease the frustration? I'm already nervous, understanding how all of these inconveniences can affect not only her but also me. We each tried everything to make a hard day better. I did my best not to make eye contact, but I *couldn't* help myself. I also knew what kind of train wreck was about to occur. I knew she was already annoyed just to have to go to work, let alone not have power in our house. I'm pissed at this scenario also. Needless to say, she was not in a happy mood over the fact that there is no electricity. And that is understandable. She was almost ready to go off to work her shift. In addition, we were attempting to eat some kind of meal before leaving. It is most difficult to grab a decent meal with no power. It was a very rough day!

I drove my wife to work. I said goodbye, telling her that I'd let her know when the power was back on, and that I would pick her up in the morning after her shift was over. The electricity goes out 3-4 times a year in the little corner of our housing development where we live. I think it's a conspiracy to irritate the neighbors. Some remaining areas of our town had full electrical power service, except for our small black hole domain where all the energy must be sucked away into an unknown void. This seems to happen for reasons beyond my vast intellect. The manner in which we lose power in our small corner of the neighborhood could be a summer thunderstorm, a snowstorm in winter, or a family of squirrels farting in the wind too close to an electrical transformer.

So, as soon as I returned home from my taxi duty, I made myself a stiff vodka martini (shaken, not stirred – James Bond style). Wonderful, now I can revel in the darkness of the house. My three golden retrievers were all wide-eyed with nervous apprehension as I frantically and loudly looked for ice for my drink. My frustration now is being taken out on the ice maker. A couple of candles and a very weak flashlight in the process can make for an annoying, arduous, non-adventurous journey toward irritating your last nerve. No power to get onto the computer because the internet is down, can't read by candlelight, and no idiotic TV to waste my mind.

My daughter from the next town over called to see how I was doing during the storm. I'm thinking, do you really want to know? I did appreciate her phone call. However, sharing one's thoughts and feelings, good or bad, among family is what it's all about. And her concern was much appreciated. The truth is, I was sitting alone in the dark with three dogs by my feet while drinking my *second* vodka martini already. By now, the unrecognizable shadows from several candles flickering on the walls even made the dogs' hair stand on end. All I had was a very dim flashlight and candles. The dogs were wide-eyed as they all gazed at me with bewilderment. Each one was nervously looking around, and all three were lying by my feet as I had my samurai sword across my lap. We all know that monsters come out in the dark. So, I was ready for them with a martini in one hand and a sword in the other. There just seemed something physiologically wrong with this scenario. Am I in some kind of horror film without my knowledge?

I told my daughter on the phone that I dropped mom off at work, which was fine because maybe the change in environment might make a tough day a little better. We all get under each other's skin sometimes, but it's often over meaningless or pointless things and misunderstandings, which arise to their fullest height. Patience is not one of the strongest qualities of humans. However, we do what we can and make the best out of difficult circumstances. Occasionally, I believe it is harder to keep one's composure as an adult than a child.

Perhaps it was just the combination of the electrical power being out and the chaotic dinner we both endured. It is never any fun being stuck in the house all day with foul rain and winds. Even our female golden retriever refused to go out and pee in the backyard. I admit there was a brief second of gleeful optimism when the electricity blinked 2-3 times, which I thought might mean we were lucky. But that was only for a millisecond. It then stayed out for another seven hours. There was nothing more to do. I finished my *third* vodka martini. Then, I decided to take the three big dogs, with my dim flashlight, upstairs to go to bed. It's about cause and effect. We cannot seem to control either. Our occasional frustrations and impatience show themselves from time to time. People's behavior is a response to events that continue to occur randomly. And we react in the manner that we do because we are only human.

We all have our moments of struggle and attempt to make sense of issues that confront us. This includes those who are very young, as well as the experienced and mature person. Imagine how it must have been for humans thousands of years ago to rationalize their rapidly changing world. Perhaps their use of more instinctual responses simplified their actions and mental stability as opposed to over-examining issues as we sometimes do today. We may not need to look too far to make educated guesses about our ancestors' behaviors. Simply look at young children's approaches in how they respond to their own little world today. It is basic, simple, and often right on point. At the very least, we can clearly see an incredible amount of humor in what they do so very well. As they grow to become mature adults, one can only hope they acquire and recognize their own strengths for endurance and weaknesses to improve. We need to break old habits.

People of different nations around the world are connected by a common thread called humanity. The common characteristics are undeniable. Due to unique cultural differences, I don't believe most people around the world would be so upset by power outages as Americans. Maybe it's living in the Northeast area where tensions appear quite high. Just drive in a car anywhere in the Northeast US.

Very aggressive and in a hurry for no real reason. I have seen enough road rage in one day to know this behavior is inexcusable. It occasionally seems that not everyone realizes what that phrase of humanity truly means. The term "human" is the root of the word. We are only human and are all connected. In the end, nothing separates us in life. Are your actions *that* important? And to whom? Remember, we are of one species. Success, wealth, and public status often segregate people into discriminatory classes. As they say, we must have the right people for the right job. *True!* But hard work, determination, and never giving up can give you the potential to reach those dreams. However, this can be accomplished in a respectful manner, not by "*class.*" That's insulting. We have the ability to do so much for one another. How rarely do we realize when we're helping people in ways we'll never know. The same is true for how we may harm others without our knowing. Therefore, a significant question is: when did the actions of our ancestors begin for the sake of humanity? Or are we still working on this?

It may have begun in small, simple ways with little steps toward helping each member of the tribe be successful. That is the first step to begin assisting the group. We all have a role to play. Although rarely do we know what that part is. As we age and mature, we begin to believe what our purpose is. But with many, *doubt* sets in, and we can never really know. It is impossible to predict anything regarding the possible outcome of an event. Some things may be immediately apparent, but the long-term consequences are often shrouded by a changing mist or hidden in shadow. The smallest act that you may inadvertently do could easily lead to the greatest accomplishment by someone else. It is all about your service and how you may unknowingly influence others. And somewhere along our evolutionary path, the act of service became evident in its true value.

* * *

The notable Samurai of Japan many years ago were feared and respected for their physical prowess. They had a strong ethical code to their masters, highly respectful and loyal. Samurai warriors had unwavering self-discipline toward their moral beliefs and code of conduct. Their ethical code of bushido may have appeared violent and harsh. However, they remained loyal to the Emperor and powerful territory owners whom they served. Samurai, or the "way of the warrior," means to serve. It represents the acts of retribution, courage, benevolence, respect, honesty, honor, and loyalty. These are qualities that one should dedicate their entire life to upholding and following through.

The way of the warrior is noble and often unobtainable by most. The samurai were not self-centered savages who dominated people but provided structure, order, and obedience from subordinates. It may be hard to understand or accept, but order, structure, and discipline are necessary in any civil society. Rules must be obeyed, and consequences accepted for violating such policies. It appears to be the manner in which the rules are determined and consequences enforced that creates contention among different societies or groups of people. There are just too many of us to come to an agreeable set of the numerous legislated laws, consequences, and ordinances. Not everyone is going to be happy. Not that I care. However, in a fairly run democracy, voices can be heard, and votes cast to incorporate or amend such rulings for the people. I wish that I could say that current politics operated that smoothly. There are always power and special interest issues that many political leaders personally seek with very minimal intent to serve the people they represent. It's all about them. And that is unforgivable.

In the animal world, the leaders would be recognized as alpha males and alpha females. An individual within every animal group needs to accept that enormous responsibility. It is often genetically driven and meant to ensure the well-being of the pack. Human nature is driven by structure and order. Humans seek this out in the unpredictable chaos of life that is bound to a disorderly universe. Leadership behavior or intellectual guidance that humans possess is multi-layered and complex but necessary. Keep in mind that we are still animals. Nature has a way of making things happen whether we want it that way or not. However, there are true leaders and those who are not. False leadership is limited to *the human animal.* With us, genes alone may not be at work, but rather *emotions* that drive our actions.

Even during the ruthless period through human history around the world, there are those that serve. It doesn't always need to be so glorious or honorable, but serving others is inbred in our nature. Although we have this unusual need to correct forms of injustice, to right a wrong, and sacrifice ourselves for others, there is still this complex desire for personal gain, which has been a mental conflict and personal battle within our entire human history. We cannot all be alpha males and alpha females. Subordinates serve a valuable purpose to the tribe. Without one, the other could not exist. If not, the tribe diminishes.

This structure of order, not like the Samurai but functioning "*family*," began at a point several thousand years ago. And there were many unjust actions along with our civility. Unfortunately, the human path of our ancestral evolution would embrace vanity and greed. Those negative qualities within our nature must be given up. This destroys all sense of organized living within a group. If we are to truly deserve harmonic survival, there must be leaders and respected workers. There must be winners and losers. Competition is part of animal nature. Emotions have no role in determining who succeeds and who fails. That is just life. Humankind must play by the rules of nature and check selfish emotions at the door. Real strength relies on doing what is right for the tribe and still allows for personal ambitions for success. There are going to be winners and losers, and that is the end of it! However, to survive as we must, true dedication to humans through acts of kindness and service is to be an active part of human nature. Humans need to see their weaknesses before they can begin to express their strengths. We should all live as the "Way of the Warrior," confronting injustice and serving *all life* in the hope that we ourselves, in turn, will someday be saved by those with similar morals.

People continue to seek a better life through truth about their objective desires and ambitions. The approach plays a significant role in whether one believes life is getting better for them. Fool me once, only. Never fool me twice. But when it comes to someone's idea of the better life concept, they seem to be fooled continuously. This may be because few actually seem to know what they truly possess. The avoidance of telling others the truth or lying almost appears to be a prerequisite for some. It is either those ranging from the average individual trying to make a living to big business and politics. There are those that believe in the idea that the common folk don't need to know all the collected facts. It may negatively impact their emotional reactions or lead to even deeper questions about an event or issue. Oh, let's not ask a question that a "leader" cannot answer. Do not forget, but always remember, the wizard beyond the curtain is not all-powerful. We have now evolved into having a "*ruling*" class which determines how we are to think and how we are to behave. Why? For the better good? I doubt it! This all comes down to the abuse of *power.* And you are the recipient.

As common folk listen to what they're being informed about local and world events, our only course is to seek further questions as to what we're being told. This fuels fear among those who work feverishly in controlling society. As Jack Nicholson states in *A Few Good Men,* "You want the truth? You can't handle the truth." That being said, I do have some understanding as to why that may be necessary. It is not that human nature is to doubt stories, news reports, or what someone tells us. It may be to avoid conspiracy concepts which are destructive. Do you ever think that we'll ever receive reliable, honest sources of information? I believe most people now know that we shouldn't believe everything that we're told. But to have trust, we depend on noble people to report news who seek out honest answers to honest questions pertaining to local and world events. As the evening news and newspapers may direct information toward satisfying and easing tension among the masses, human nature forces us to demand and dig down to reach for real answers to questions we all have. We claim that we want the truth; however, we also want to hear that which makes us *feel* most comfortable.

* * *

For humans, it seems that it is all about what we want. Nonetheless, truth or not, we want what we want, whether we understand it or question its true value. I find this the most hilarious characteristic about humans. I don't even know how to fit it into a category within human nature. Observing behaviors today may not always be considered our finest hour. It will probably be questioned when we look back on our past. Especially if what we want becomes more and more outrageous in obtaining those answers. Fortunately, science is dependent upon asking questions and testing hypotheses. Science in our time will never have all of the answers you are looking for. But it is a good start. Science is there to be proven wrong. And that is what further drives our exploration into a search for a momentary truth. If, however, you are seeking answers from higher political powers, good luck. There are just some things you're not meant to know.

When you get right down to it, people living almost anywhere want just a handful of things. Some are long-term goals or perhaps daily vices. Such as, I couldn't begin my day without coffee, and it's not for the caffeine, but the aroma and taste. It is much appreciated. We could discuss these things we want in general terms or more specific details. I'll speak from my point of view and observations made about myself and others, including friends, family, and those strange, uniquely different acquaintances I've met during life. Either way, it generally makes me laugh as to what makes humans happy. It's like a kid in a candy store. When is too much chocolate too much chocolate? OK, that's not a good example, there's never too much chocolate.

Human behavior ranges from simple cravings to the deepest dreams coming to fruition. People who don't have enough money in their wallets to buy a decent meal will spend it all on tickets when the jackpot lottery reaches 700 or 800 million dollars. And every one of those ticket purchasers believes that they alone have the coveted winning sequence of lottery numbers for the drawing. Even I have a secret system for hitting the lottery. I just keep telling my wife that I'm refining that system each week. And there will be times when you know for sure *this* is the week. The day after the drawing, when I gaze into the bathroom mirror, I see a jackass looking back at me. But optimism and ego to beat the system have me continually working the numbers.

Another interesting human behavior is the occasional indulgence of junk food obsessions. We have those who are secret hoarders of junk food and those who blatantly go out to their local burrito shack and buy a large cheap bag of garbage for $2.99. Afterwards, you may regret those impulses but do it anyway. I'll never tell anyone if I have mini chocolate-covered Easter eggs for myself, strategically hidden throughout the house. But if accused by your wife of food that is gone, blame it on the kids for eating the last of the snacks. She won't get angry at them. If all else fails, *it was always the dog.*

It is also human nature to wait until the last minute to do something. My wife will say, "I think the kitchen trash needs to be emptied." I just stomp my foot on it and say, "We'll get at least another day out of this one." The best move is when someone in our household would empty the upstairs bathroom trash, tie the small bag into a knot, bring it downstairs, go through the laundry room, and leave it on the doorknob that leads outside. All anyone needed to do was open the door, walk ten feet, and place the bag into the trash can. However, maybe that's my job. In a way, I can just hear it now: Neanderthal wives saying, "Honey, take these bison bones out of this cave right now." And he'd reply, "Yes, dear." And that's how that *phrase* was invented! But when it comes to filing income taxes, wait until the last week of the April deadline. Your accountant hates it, but I don't want to part with a dime any sooner than I must. Fortunately, my accountant is one of my best friends and understands me very well.

Unfortunately, being only human is often to criticize or discriminate against someone because they don't agree or look like you. How primitive! There is zero room for discrimination anywhere. You may disagree, argue, or ignore behavior, but don't harm others with words. It's unjust and not necessary. The 1997 science fiction movie **GATTACA** is a film about a futuristic society that seeks genetic perfection in human society. The film starred Ethan Hawke, Jude Law, and Uma Thurman, and several other fine actors. There are two main classes of people in the movie: those who have been genetically altered by having the sperm and egg of the parents prior to conception manipulated to eliminate predispositions for genetically physical and emotional imperfections. This may include such disorders as cardiovascular disease, diabetes, obesity, compulsive disorders, and intellectual challenges that could limit one's optimum potential performance in society. They still receive the qualities of the parent, but only the best traits from each. It is a noble act of science and technology.

Then there are those individuals who are born the "*natural*" way with unknown genetic imperfections that may be passed on from each parent as it occurs today. Throughout human history, the natural way for creating new generations is that fifty percent of the genetic material (DNA) is given to the offspring by mom and the other unknown genetic material half by dad. You never know what the offspring is about to receive or which traits you give as a parent. As a result, we are the direct genetic combination of our parents, which includes the "*good*" traits and those which may be considered "*bad*" or perhaps those of a lesser advantage for the offspring. Currently, genetic variation by chance is nature's way.

Genetically engineered individuals in the movie **GATTACA** are viewed as the elite group of society's people, with preferences toward superior employment and status within the community. Anyone not genetically modified prior to conception to eliminate those predispositions for disease, illnesses, or inferior abilities is considered unworthy of any status among society. Good or bad, this is how we've been coming into this world form for eons. The resulting two separate classes of people created in the movie now show a whole new distinct view of discrimination. It is now elevated to an entirely new level. Discrimination in the *film* is no longer based upon visual or behavioral manners against one another, but rather on the quality of their genetic profile.

Since the beginning of time, people have sized up one another to see how they fit into the tribe. In many animal groups, a dominant male and dominant female control the pack, keeping a sense of order for the strength and safety of the group. They are the breeders. Challengers jockey for position at becoming the dominant "leader" of the group. This is nature's way of ensuring that the best genes remain within the pack. Rarely do dominant challenges fight to the death. But the strong ensure that their breeding qualities or traits will create new generations of offspring with the best chance to continue strength and stability within the group. This is not exactly nature's way for people. It is all by random chance in whom we choose to engage in creating our offspring. In the movie, this can be eliminated through genetic research and application of such information.

The concept of **GATTACA** has a sound objective. The use of technology to eliminate predispositions for diseases or disorders that might ordinarily be passed onto offspring as it can occur today would be a welcomed accomplishment. Who wouldn't want this? However, if some parents were not able to see their local geneticist, as in the movie **GATTACA**, to modify their newly created offspring by removing "genetic defects" or imperfections, they would be segregated as "flawed." Theoretically, these individuals would not be able to compete with those with superior genetic qualities. Discrimination at this point would now be based upon the level of one's genetic profile. It is something that is very different than what we have today.

I do have grave concerns with humans and their behavior to prove superiority in the field of science. Egos and greed are massive in the science community. It stems from a potential hubris attitude in an event to close the current gap for a universal acceptance in us to conduct gene alterations in humans for the "*good*" of our species. That which initially begins as a noble quest often changes direction for personal gain and recognition, and financial benefit, even in the world of science. And that power is dangerous without strict moral limitations. I'm not entirely convinced we have that dedication for the better good. I so much wish we did, but this is the flaw of human nature.

Even now, new scientific discoveries are copyrighted as "ownership rights." Any additional scientist wishing to utilize this discovered data must now pay for rights to further conduct research. *So much for the better good.* Greed has now entered the scientific community. I suppose there is no such noble act for the sake of mankind. Regardless of initial costs spent on research to be regained, it is an absolute disgrace and embarrassment of such greed by our own kind. Take whatever credit you believe to have deserved but share the wealth of knowledge. Science was groomed to discover, learn, and help in mankind's ventures. Science cannot, should not, and will not be permitted for someone's personal gains. True science is a gift to humanity.

Currently, we genetically alter foods and other living creatures. I would give anything to reduce or limit human disease or disorders through genetic alteration if and when perfected. However, the fallout exceeding the boundaries of what we should or shouldn't do to our species should be left to the movies. I'd like to see the movie **GATTACA** have a reminding influence in how and why we proceed with human endeavors for "*advancement.*" Perhaps there are certain things that should be left alone, considering how human nature often spins good intentions to profit and personal gain. Curing illnesses and diseases is a noble cause. Altering human genes may just be stretching those boundaries. On the other hand, if we properly proceed with moral conviction towards a goal for the good of humankind, then our knowledge, experiences, and abilities should be used to their greatest end. And if you're wondering about this point of view regarding my doubt or hesitation, it is not one of a religious or spiritual nature but rather from the devastating possibility of a result within science that could go horribly wrong. And since we are only human, ulterior motives from initial good intentions for personal gain are a huge part of our human nature. Perhaps I'm less trusting or more of a cynical individual than previously thought. The funny thing is, I'm certainly not alone in those beliefs.

The human species walks a fine line between a destructive end and one with amazing altruistic qualities for success. The knowledge from our achievements through technology and creative innovation has the potential to lead us down dangerous paths. If our hubristic nature drives motives to abuse and inappropriately use the tools and weapons of the future, our time may be limited. Yet, utilized in the proper way, everything that we've gained can strengthen our resolve to overcome our differences. And there remains that delicate thin line that we frequently walk. *It is a blessing and a curse about our nature.* The main thing to never forget about human nature and human behavior is that *we are only human.* Perhaps that is an obstacle we need to hurdle and to redefine the term human.

To ask the question of what it means to be human can open generations of closed vaults about our nature. Each question we could ask about ourselves may have a different response on similar but forever expanding subjects. I don't always seek answers about unknown topics, but rather better questions to ask. Many can respond to cause-and-effect issues. But I have always been interested in the "*why*" of it all. Why do things happen the way they do? Why does an event happen when theoretically or logically it shouldn't? Those questions may be the plague for human nature. Therefore, questions such as "*why*" and answers of "how" things are accomplished will forever haunt us.

Our ancestors, somewhere around 300,000 years ago between *H. heidelbergensis*, Neanderthals, and modern humans, looked up at the night sky and began to wonder about the world around them. Their thought processes weren't just about instinct and the need to survive but may have contemplated specific events, and possibly their future. They were most likely the first to ask "why." All of my life, I've been asking "why" when others have been asking "how." I do understand that a combination of both concepts is necessary for us to advance in our development. Maybe it is our strengths balanced out by our weaknesses that make us human. On the other hand, perhaps it is conviction combined with trust in ourselves that keeps us on the game board of how to learn to become more human.

*   *   *

Humans, over a millennium, have demonstrated strength and honor. It is almost exemplified as being innate. When diplomacy is exhausted, we will fight for what we believe in. Although many exhibit this strength due to security of country, I get it and agree! Personal conflicts, however, should be resolved in a civil manner. People demonstrate strength by standing their ground. And they should. What is mine is mine, and what is yours is yours. Dogs pee on trees and fire hydrants for a reason. They are claiming territory. They are claiming this territory as their own. It's common among many animal species. And we, as humans, are no different. Although, I have yet to see politicians lift their leg upon anything. As tough as we think that we may have it, most people don't fully understand that their *worst* day here in the land of the free does not compare to the *best* days in other lands under siege by war and destruction. People become complacent when they really don't experience daily pain and suffering as many do on foreign soils. The goal and main objective of our military presence around the world is to protect our country and enable an opportunity for basic humane freedoms for innocent people. There are always peripheral objectives in the interest of one's national security or political advantages. Strength and honor require great inner discipline to stand against injustice and conduct actions in a noble manner.

Honor often requires a type of service to show one's strength. There are many forms of service which we take part in daily. Service of duty comes in the forms of either kindness, outward respect, or simple gestures to show honor. Many often confuse true strength in an individual. And that's OK too. They just haven't learned yet. The point of the conversation is: how do you recognize strength and honor in yourself and others? We all have these two qualities but often are hesitant to express them. Honor always comes first. Always remember the way of the warrior. Strength comes from within. And at times, strength may even mean to be still or silent. These two terms may be hard to know how to show. Keep in mind respect and service. Honor and strength will follow. Disregarding the facts of expressing our public opinions and freedom of speech, the varied views of the many sometimes cloud how and where we stand united. This must be through honorable strength. At the very least, we should have respect for those who may oppose us. We've agreed upon these rights for American citizens many years ago. We've fought against each other and died for our beliefs. The laws and multitudes of freedoms that have been formulated have now made our country unique to the world. Where else can two people openly oppose each other's views or beliefs with the given right to do so? However, it does come with a cost. We often lose the sense of our united front as one nation because of our inability to control those instinctual human *emotions*. This is not due to a lack of intelligence but a lack of respect for one another. It is tough to be strong and steadfast. However, that is an expression of true strength.

I appreciate people's expression of service. For example, I've driven through our neighborhood during a hot summer day, seeing a lemonade stand by the street managed by five- and six-year-old kids. I know they're trying to make a few bucks, but they did provide a service of need. I'll stop, buy a paper cup full of lemonade, and watch how delighted they are to make a sale. Of course, I never drink it. I'm sure if the local food inspector didn't come by and examine how it was made, I can only imagine. What better way to mix up a pitcher of lemonade than to stir it up with your bare hands? They're only kids having a little fun. Nonetheless, a noble expression of service. And that is admirable in many ways.

Currently, the act of service is being widely acknowledged. Service often refers to people in the medical field, security personnel such as police officers, teachers, or military personnel. And there are so many more examples. Individual warriors of the military often do so for the sake of honor and to protect the homeland. I certainly applaud that. But war or physical conflicts are an ugly beast. There generally isn't anything good that can come from war among our species. There will always be a downside. However, we will and should fight to defend our freedoms and assist in the liberties of others around the world. We are, after all, of the *same* people and of the *same* species. Conflicts and battles have become a trend of human behavior that goes back beyond recorded history. People will innately defend what they believe can be taken by others through any means available to them. It is part of our nature which we have not evolved from our past. There is an element of self-preservation. I do believe all measures should be explored before the taking of a precious life of our kin. Knowing this, it still is not easy for us to quickly act, using our worst aggressions of violence when needing to protect the greater good. This notion of the "*greater good*" can be disputed among the segregated nations around the world, as each may feel that they know what is right or good for their people. Which is true! So, what qualifies the standard as the "*greater good*"? Why is it so difficult for us to get along? Perhaps, it is human greed, power, and lack of acceptance that are the reasons. A self-destructive mode of our own kind serves no one. Those thousands of souls who were lost in numerous battles will never be forgotten. As a nation, we must have unity, for as they say, live together or die alone.

The act of service is just that! Giving to help others. It is not simply a *kind* thing to do, but an altruistic behavior and expectation to assist our own. For humans, service is personal. Many who fall into this category do not seek self-recognition. But I thought all people wish to be recognized? So true. However, this is different. Service to humanity and country goes well beyond oneself. It is close to instinctual for us to help others in need. In doing so, we protect that which is sacred and meaningful to the masses. Service should not go unrecognized but at least acknowledged as an offering of unconditional gratitude. Service is primary!

Animals will battle and have conflicts of their own kind for food, space, mates, and territories. It is instinctual and embedded into their genetic makeup. However, animals rarely fight one another to the death. The weaker of the two will leave the fight, and the stronger animal will take command of the group, herd, or pride. The alpha male and female will dominate their small contingent of members. People also battle one another; however, we will fight each other to the death. A tremendous flaw in our genetic makeup. The resulting conflict on large scales is often considered wars. How *tragic* for humans. We do not always recognize the real troubles of those living within the borders of war. The noble warriors who defend our rights for the sake of freedom and independence are heroes that can never be thanked nor appreciated enough by those so fortunate to not have to witness the suffering and brutality that humans are capable of doing to one another. Whatever the cause or reason, it is inexcusable to degrade, suppress, or destroy innocent life. It is *that* which we will fight for.

Warriors who go to battle to save or fight *against* immoral acts toward our own kin are true heroes. But they are heroes for the right reason. On one hand, we are the only type of animal life on earth that executes such devastation toward one's own kind. On the other hand, we need the ability of *acceptance*. Respecting other people for who they are. As brutal as it may be to take life, sometimes the lesser of two evils is necessary. Often, we are required to break apart from those inhumane minority groups who persecute the innocent. There is an overwhelming majority of humans who do not wish to condemn different philosophical beliefs of others. However, there is a forever line in the sand that noble people recognize that may call us to action. We all need to live and worship freely without rebuttal from those who think, believe, and act differently. The only true moral and ethical approach to one another is to have honest respect and to worship *life* itself. Honoring *all living* things, human or nonhuman, is the ultimate duty and true value of a person's worth. And that nonhuman element happens to be the earth on which we all live. As sacred as I believe that life is, I still have the utmost respect for our military personnel. They've proven their dedication and resolve to the cause and faith of the decisions made by our leaders, which at times must act against brothers in foreign lands. It is not easy for trained personnel to follow orders that bring harm to others. Yet we are adamant to keep harm free for the innocent. One of the sad components of war is labeling anyone as the *enemy*.

Perhaps, as we evolve, the term "enemy" may vanish. Yet why any radical group chooses to kill innocent people simply because of one's nationality or beliefs is a concept that I cannot understand. I only wish that someday we could emotionally rise above the misguided understanding of terms such as enemy, right and wrong, and place in the forefront words such as respect, honor, and value for living things, which includes mankind and all other living creatures. A dream, perhaps. However, I believe it is in our ability.

It is a sad day when brother must combat brother. However, the dedication and honorable stances that we make are to admire and commend our veterans who have served us. War heroes and veteran patriots are honored in our country during one single day of the year: Veteran's Day. They should be acknowledged *daily*. But they are also recognized and remembered with loving gratitude for their service during Memorial Day. Memorials, pins, medals, parades, service recognition, a warm welcome handshake, or heartfelt thank you can never completely justify the honor and respect for their duty and sacrifice that is deserved. They need to be acknowledged and appreciated more often for their service, and equally important, their choice to act in a moral and ethical manner for the good of the many, regardless of the intent of the few.

As a master swimmer, I know how my body responds to intense training and swimming. I could go to the beach, jump in the ocean, and begin swimming as far as I can. Once drastic fatigue sets in, I would then turn around and swim back to shore. However, if I were to continue swimming eastward out to sea, I would reach a critical point of no return. I would simply run out of energy to make it back. This would result in a serious consequence that I could not survive. It is all about a survival point. We must master the mental strength of actions against others. It is not just about the survival of the fittest. But survival for *mental* challenges. These may place us in serious danger and great harm as a species.

After endless physical conflicts and wars in our human history, we are reaching a "*critical point.*" Fall after fall of countless human empires are somehow replaced by another. Without calling this innate behavior, it almost appears as though there is somewhere a suicide gene embedded into our genome. Eventually, we as a species may someday reach a point of no return. War of any extreme is almost instilled in us all, no matter how peaceful we can be. I find studying evolution an interesting subject because life and death are in the balance. It indicates change in living things over time. It also can show the possible end to a species under specific circumstances. Unfortunately, extinction is forever! And I understand that conflict is often necessary for survival. That means some will live, while others will die. The frightening thing is it can happen suddenly to any living creature, including us. I only hope that we are not approaching that critical point for our species. "Oh, no, no," say some. "Not in my day." But who can really say? The point of no return is final. And I *do* know how quickly one can run out of gas swimming.

All animal species, including us, are a product of our parents. It's astounding how their qualities are given to their offspring. Good or bad, right or wrong, we inherit much of what we are. We become individuals in our own way, but the family tree is not far behind. There are many leaves on that tree. Each slightly different and similar in nature. Parental guidance and examples of their lives often "rub off" on us. In the end, customs, values, and traditions arise. For some, it may be a curse; for others, a pure blessing. We cannot explain how humans came to be or where we were before our lives emerged. This is a total mystery. The same is true for where we are going after this short, precise life. Some things are just meant to remain unknown. The best that we may have is *faith.* The idea of trust in a belief is more powerful than hopeless suppositions of those unknowns. And I'm OK with that. It's your life! Make your choice.

*   *   *

There are billions of people on earth today. That being said, there are billions of separate worlds among our population. Everyone has their own understanding of who they are, what they believe in, and perceived destiny for themselves. It is impossible for anyone to perceive what existed for a life prior to birth. Perhaps there are billions of concepts. As we live this life, one can only speculate about what comes afterward. Human nature is so difficult to precisely identify since there are a billion different worlds. Who are we to criticize what others believe? Each one is unique to the individual. There is a laundry list of human qualities to consider. Forgiveness, honor, and respect may seem to be the three main ingredients on that list. But there are so many others. Our ancient ancestors had to develop codes of conduct that aided all individuals of their tribe. Eventually, these codes evolved into the numerous qualities of current humans. This can easily lead to some contentions among us. Nonetheless, we are social animals and must get along. Whether you believe someone is within your world or not doesn't matter. We are gifted with our own personal views of it.

Many creatures in the animal world also have a social network and hierarchy of behaviors. There are dominant and subordinate members among many species. These are not particularly favorable terms to use when it comes to humans. Who is in charge, or making decisions in a human family of four or five individuals? I know, I just threw a hand grenade into a room and walked away. Sorry about that! Often, these tasks are designated equally to both mom and dad. But how often do you hear dad say to his kids, "Go ask your mother"? Mom will say, "Wait until your father gets home"! To live a human life is difficult. Strength, discipline, and maintaining harmony are often an arduous task, especially when mom and dad are not on the same page. There is loyalty among mates. Loyalty displays respect for others' beliefs and actions. The level of respect is one of the most important aspects for humans to peacefully coexist. Some people may know perfectly well that loyalty can waver. Support one another, even among differences. It's not too much to ask. It is an expectation.

Since we are only human, there are so many qualities that make us who we are. And that list seems to continue to grow. But another important consideration is neatness. In our home, the last person who sleeps in and gets out of bed last must make the bed. Easy, right? Well, you're half asleep and that king-size bed has a sheet first, then a light blanket, followed by two comforters and four pillows to arrange. OK, that's still not difficult. I'm often the last out of bed. But make sure to align the shoes in the master walk-in closet. I step on them every time I go inside. Oh, I know, most of them are my shoes. But that's alright. My wife usually doesn't mind my laziness. I'm not going to complain about my own faults. I'll do similar things around the house also. Nonetheless, I believe one should do something hard every day. It could be a small thing which you may not feel like attempting at that moment, or simply putting off a chore until tomorrow which should have been done today. We've all heard that story. Succeeding in a challenge strengthens your personality. I'm not implying to do something so difficult as in Special Warfare Operations training as in BUDs, Navy SEAL Hell Week training. Just something which improves your discipline. After all, how many people can be a tier one operator?

During family gatherings or when friends come over, my wife cleans as if the Pope were arriving. I run the vacuum on the family room carpet, the oriental rugs in the hall, dining room, and den. One thing that I've been asked to do is push the vacuum on the living room carpet so the "*lines*" are all straight and there is a uniform pattern. Everyone will then know that the carpet has been cleaned. Hell, I could just push the vacuum quickly to get those lines without vacuuming. But I don't mind doing the necessary things around the house to keep it tidy. However, I'm really not the best about keeping everything clean. My wife is exceptional at that. So, I don't mind running the vacuum or cleaning floors. After I finish, she will say, "Wait, I need to dust first. We have plenty of dust," I reply. She's not amused. And rightfully so. My wife and I are very good when it comes to entertaining, making sure people have enough to eat and drink, and making everyone feel very comfortable when they come over. She is really fantastic, but no one can get ten pounds of shit in a five-pound bag better than her. After being exhausted from cleaning and buying food and drink for our guests, she will ask if I would get some touch-up paint for spots in the family room. "What? People are coming over in less than two hours!" Now that would be eleven pounds of shit. Not happening. Finally, all goes well as planned and everyone has a great time as we fill our bellies and tell our tales. We typically poke fun at each other as to what we think is important.

I believe that most humans attempt to be the best that they can. It is our nature to want to succeed and be recognized. Making progress is fine. However, keeping some things the same is good, especially traditions. This is exceptionally true when you have a gaggle of relatives coming over. We have enough challenges today. In fact, it has always been that way over our evolution. Every day can be a battle. Somehow, we get through it and the next day we continue fighting. Other times certain things just can't be overcome. These are lessons which we must learn or adjust to. No matter how mentally or physically strong you may be, some challenges are just that. Therefore, dig deep and attempt to use your skills for defeating those obstacles. You are stronger than you realize. And being strong will tell you when to let it go. There's nothing wrong with that. Remember, we are only human.

To keep humans on the game board, there are specific issues which we must recognize that make us who we are. Some may be acceptable and just. Others, not so much. But with every good there will certainly be evil to combat. People must have conviction about what we wish to have, and what we are willing to accept for humankind. Humans must demonstrate a strong resolve for their moral issues and stand firm against those without noble intentions. If not, all may be lost. If the game board appears incomplete, add more pieces. Those indistinguishable pieces may simply be accepting the fact that you will never win the game. Or, perhaps, give you the edge to outcompete your foe. Unfortunately, life is not a child's game. Nonetheless, we appear to move through life as game pieces. Just go with it. Keep a level head, and life will magically work itself out. However, there is always the need to be resolved in our undertakings.

Humans have the opportunity for great strength to implement compassion, honor, and loyalty. We are close to the ability to conduct moral and ethical actions toward all humans and honor all living creatures. It is a matter of time until we collectively understand that we need to survive as a species. *Again, it is all a matter of time.* Time in which we use wisely or waste toward total failure. Free will and the choice to decide our future rely on everyone, possessing clear moral actions. How little do we realize that we need each other to survive.

As I have indicated, I am not an authority on what it means to be human. We can see that there are so many different aspects which fit these criteria. To list just a few would be an injustice. Nonetheless, different individuals will have different opinions on this subject. And that's OK. Here are just a few that I've recognized.

We are aggressive toward each other, verbally and physically under certain circumstances. We fight one another to the death. We also start or finish wars with our own kin. People are destructive to our earth. We seek power over others. Power seeks more power. When we age, we sometimes have the emotional reactions of an infant. We have the desire for greed, which exists in many forms. We fear that which we do not understand. Our primitive instincts are to fight or flee situations that are unsettling. Rationalization is a higher evolutionary achievement. And the last physical development of the human brain is the ability of judgment. We have the potential to do great things. We have been designed to do so. The main obstacle is keeping a conscious check on our evolving emotions.

People are generally empathic and may risk their own lives to save others. We help one another in desperate times of need. We are at our best when things are at their worst. People seek unconditional love from others. It is fortunate then that we have dogs. People look for a mate to share our lives. In return, we pick up others when it is needed. We have incredible strength of resolve and creative abilities which only humankind can possess. People can work through problems with logical reasoning and using past experiences. And there are things in life which will never be forgotten. In days of despair, many of us will never give up. We are all these things and more because we are only human.

Wars will diminish, and all life will be viewed as sacred, *but not today.* The earth will heal, and waters will run clean, *but not today.* People will respect each human as their own kin, *but not today.* Greed will fade, *but not today.* The world will break from darkness and light will shine, *but not today.* Beasts will live free from mankind and dwell in natural environments, *but not today.* Fear will fade, and happiness will return, *but not today.* Humankind will become more forgiving and protect all life on earth, *but not today.* We have so much to overcome. Because we are only human.

Chapter 10: It Is Time

The thing we take for granted, beyond the companionship of other people, is time. When we were young, time moved slowly onward. Our awareness of time is not only limited, but it moves faster than we would like as we grow older. And the timing of your life can be taken away without anyone seeing it approaching or ever expecting it to happen. Although time is constant, it seems to slow when horrific events or tragedy occur. I can recall this when hearing of family members who pass. It was clear to me that when my father passed away, we did not see it coming. That phone call in the middle of the night, telling us to rush to the hospital, shocked both my wife and me. Everything around me felt as though I was in slow motion at the mention of my father that night. Time is one thing that can stop instantly for some, without warning.

What humans ultimately want is more time. We rarely think of our survival timeline until something out of the blue comes along. For the more time we have, the longer we survive. The list of human desires we seem to have in common is endless. Most men are the same, and women are all uniquely similar. There are differences between us, however, and that is expected and natural.

We have these uncanny common behaviors, and time is one of them. Fortunately, the items which I have previously mentioned are trivial, yet familiar, just as your face is in the mirror. These things about us allow me to respectfully laugh at who we believe we are. On the surface, we may see differences. Go a little deeper, and we will recognize similarities. Ultimately, we are all the same. If you are able to find the core of an individual, you will know immediately how much of a companion they can be. And time is what we want. But true companions are like dogs, they last a lifetime. We all want more time.

We all should make the most of our given time. Doing everything one could possibly achieve means nothing unless it is meaningful and has value. It is often much better to act slower, but more accurately and precisely, with the tendency for the task to have personal significance. Time should be reflected upon. Meditation, yoga, performing Tai Chi, or other mental or physical disciplines relax and filter out the mind's senseless daily input of useless rituals for a brief period. It strengthens the character of the person. It is never a waste of time when you accomplish that. This is important, to push the pause button on your life for an hour or so each day. Time then becomes more appreciated and utilized more wisely.

Physical anthropology, and the ways we believe our ancestors lived over the last several thousand years, fascinates me. I always wonder why, in addition to how, we have developed behaviorally. We can examine artifacts and sites on where and how our ancestors lived, but we cannot go into their minds and ways of thinking. That would be more speculation than anything else. People may know what they know, or of things that they do not know, but we do not know about things that are unknown. I would like to know which ancestral species first began to think about time, or the lack thereof, and when it became important to them. We all realize that time was needed to achieve basic survival needs. But was there ever the urgency for our ancestors to make the most out of time?

I hear about people's tense frustration and the sacrifices they undertake for the sake of their children. They are so busy running them from here to there for whatever the child's activity may be. Nonetheless, having enough time always appears to be a problem. First, there should be some time for yourself. If not, then you are doing far too much. Believe me, my kids and grandchildren mean everything to my wife and me, but there is still time for them and for ourselves. So, I understand people's noble intentions about giving to others.

Secondly, are you giving your child too many choices in the activities they want, which require you to shuffle them around to a point of exhaustion? They must learn sacrifice too. I understand that life really is short for what humans want out of it, but a balanced perspective must be met by both parent and child. I can hear Neanderthal dads saying to their boys, "son, you are not going anywhere until you sharpen that spear." Actually, I do not believe they ever said that, but it was probably in their minds as to what must be done. Therefore, what kinds of influences demand that we use time in the manner that we do? Is it for pure selfish reasons, totally for the needs of others, or a perfect balance which satisfies us *fairly*? You need to look at that. Then again, there are people who will do whatever is necessary for the circumstance. If so, then do not complain.

Ask most people in the country today, and they will say there just is not enough time in the day. Others who may be retired from the workforce claim they have all the time in the world. So why does time differ based upon your age? The activities? Perhaps it is what we do with our time that really matters. Often, doing too many things at once makes us feel that there is not enough time. Perhaps if we pace ourselves, or prioritize our needs, one may feel as though there is plenty of time. And *why* do we think that everything is so important? Do you think maybe, as young adults, we do too much either for ourselves or our kids?

People may want to think about drawing boundaries when it comes to time spent, especially if they are stressed out about cramming too many things into a single day. Sorry, you cannot get any more time than twenty four hours out of one day. Then there are those who fear that there will never be enough time. Slow the mind, organize, and you will find time. Just remember the country rock band *Eagles* song, "Take It Easy." And for God's sake, it is not the amount of time you have, but the quality in which you use it. I know so many people who were taken far too young. Do not run yourself into the ground.

Nothing signifies timing more than the seasonal changes. Like nature's timepiece, life adapts to yearly weather changes. From that point of view, there is order in this disorderly universe. It is almost as if this world was destined to play out life according to detailed events, like the sands flowing through an hourglass. Every grain of sand would be a living entity or event which needs to be played out. And from that proper perspective, time never runs out. When it does in an hourglass, simply turn it upside down, and time continues onward.

Our role or purpose is once again on the game board, but not always in the same game. It is a crazy way of presuming that our essence is always present, but in a different form. From my perspective, the body has an expiration date, but the soul is never far from loved ones. This is *my personal belief*, and not necessarily everyone's. Studying and teaching science for many years has given me an understanding of certain aspects about life on Earth. Personal beliefs of this life are just that, *personal*. As a biologist, I have always said that it is harder to stop life from occurring in this universe than it is to start it. However, time must take its course for any type of life to ever begin anywhere. Eventually, evolution dictates change, extinction, and who lives on. Someday, it will come to an end. As these events occur, we refer to life being played out until, "*when it is time*."

When you consider the lifespan of an average dog in comparison to the average lifespan of a human, each day to a dog does not equate to that of us. Mathematically, on average, we live seven times longer than they do. Therefore, when you consider time, each day for a dog equals one week for a human. It makes you want to do all that you can for your dog each day, for their time is so very short. For some reason, a superior power has decided that the most precious gift to mankind is to live for only a short period. Unfair, perhaps? It is a curse, and yet a blessing.

I recall speaking to another teacher who was very upset at school one day. She told me that her dog was dying and had a few days left to live. I replied, "what are you doing here? Go home! Spend every moment with your dog while you still can." At first, she seemed shocked by my statement. She then realized that I was right. I indicated to her how time passes for a dog, and not to waste it. The most important thing was for her to go home and leave the work, which will always be there. She thanked me profusely. It was the only way in which I knew that I could help. Speak the truth! Spend as much time with your loving companions. Time is *sacred*! Use it wisely.

I do understand the scientific perspective in the flow of energy and the first law of thermodynamics, which states energy cannot be created nor destroyed. This law has a very interesting concept in relation to life and the amount of time given. Yet, to speculate about an afterlife would require an entire theoretical explanation in a book all to itself. I do believe there is more to life than simply my body being on Earth for a few years, but that is a declaration and a right of my own. I am more than just a leaf on a tree, and after one season to be gone forever. There is a significant impact of every leaf on a tree. The very existence of that leaf has a purpose for itself and the well-being of its tree. The decomposition of that leaf will strengthen that tree. I exist because of energy, just like that tree. Since my energy cannot be destroyed, where will it go? A transformation to where? What will occur to me, as a person, when my time runs out? Personal beliefs, as an answer to this question, may satisfy some and petrify others. There is nothing that you can do about it anyway. Therefore, *fear nothing*!

So, when did our ancestors begin contemplating this idea that there is more to life than simply what we see in front of us? And more importantly to me, the question, *why*? Is the multitude of our emotions and actions taken beginning to make us human? We need to understand some aspects of human evolution before one can really begin to ponder these questions. Small groups of our very distant ancestors would not have had specific feelings or even loved one another thousands of years ago. Yet, their literal thinking capabilities would simply have them abandon their dead wherever they remained. No ceremony, no burial, just a dead member to be left behind. Their brains would not have had the ability to perceive beyond their literal world. Saddened, perhaps, but they were just no longer part of the tribe. They were just simply gone. We are now beginning to have a better knowledge of where we came from and how we developed over time. An emotional evolution began to eventually feel deeply for their deceased kin. The ceremony of a burial then began. Thank the Neanderthals for that. And this has a rather significant impact on those whom we leave behind now and how humans re-envision their own future without them.

An understanding of the behavioral changes that have occurred is all about timing. And when the time came for early humans to wonder beyond everyday events, creativity and expanded visions gave rise to different ways of seeing the world. It was all about language, insight, contemplation, and communication. It was seeing an innovative process and then perceiving what and how it may impact their future. This involved forward thinking, and not just for the moment at hand. They were slowly becoming more human.

From that point, thousands of characteristic splinters of independent aspects of how we live made us become human. And new visions and concepts we develop through mind and body continually remold the nature of what it means to be human. We have much strength and many weaknesses, but we are only human. We have capabilities now that can reshape the world, which indicates that our nature will always change. A duck may look like a duck, walk like a duck, and sound like a duck, but it is just a duck. Our evolution into becoming human will never be the *same* human. This is because of an unlimited diversity of behaviors and beliefs that continue to change among us. One human being is not just like another. Fortunately!

* * *

There is certainly a difference between time and timing. They say that timing is everything. I believe most everyone understands the term "timing." It is a word that can be utilized in several ways. One may discuss the *timing* of a baseball player attempting to hit a pitched ball, or perhaps when to ask another person to go out on a date. The atmosphere must be just right. The telling of tales or comedy must fit into the proper sequence of the story or punch line. We use "*timing*" every day. And then there are moments when our timing is off.

We hope that the words that are said can have the correct meaning or response for others. Timing is most important regarding a hopeful or expected outcome. Unfortunately, people often say or do the wrong things at the wrong times. This is *because* we are only human. People attempt to conduct themselves with the greatest intention, only to fail by accident. Nonetheless, we try to say and do the right things at the right times. At least I hope we do. I know all too well that I can fit either my entire left or right foot into my mouth at will. It is always unintended, but I have done it so often. Sometimes the most humorous things are said or thought after these actions are performed. And still, there are those that leave a lasting memory of our ridiculous behaviors. Accidents! Misunderstandings! Spoken words! These, and so many more references, often result from poor or inappropriate "timing."

In regard to timing, I clearly recall the day when I was at an athletic club after one of my intense swimming workouts. I trained with several members of our Master's swimming team four times a week. There was a section of the massive pool just for us to train. We were all good friends, so the strenuous workout made it bearable. Timing is important in any sport, and swimming and water survival techniques are no different. You do not want to go out too hard in a swim, but fast enough to compete. The last thing you want is to not have enough energy left to finish the race or swim challenge. There is a balanced discipline to being competitive without using all your strength too quickly. Either way, you were dead tired at the end of training. It was time to regenerate. A long soak in hot water seemed just right.

Next to the pool was a large hot tub. I thought a good soak would ease the muscles. There was a woman I had never met sitting on the edge of the hot tub. Trying to be nice, I looked at her and said, "Oh, when are you going to have your baby?" She replied unemotionally, "I already did several weeks ago!" **Oops!** It made me think twice about striking up casual conversations regarding delicate subjects. I meant no harm, just thought that she was expecting soon. I was just trying to be nice. Talk about poor timing and shoving a foot in one's mouth.

I must have gotten that skill from my mom. She was the most pleasant, caring, and thoughtful person to have ever lived. Mom would strike a casual conversation with almost anyone, whether she knew them or not. She would do this just to be nice or courteous. One day, my mom was sitting at the hairdresser, waiting for her appointment to have her hair styled. She saw a woman sitting in a chair not far from her and said, "Oh, are you next?" The woman *coldly* replied, "No, I am done!" **Oops**! I guess if you cannot tell whether someone just had their hair styled, it is a legitimate question to ask. I would have laughed, but Mom probably turned a slight shade of red.

Tales like that one could easily fill a book. But that was my mom, always intending to do the right thing and never wanting to bring anyone insult or harm. When it was her time to pass, we all felt an unforgettable loss. It is sad that there will be no more new tales to talk about Mom's life, but we continue to reminisce about those she made. Sometimes daily for me.

Timing also has a lot to do with luck. It is kind of like running around in your backyard as a child and not stepping into a pile of dog pooh. It was either luck or skill if you came away cleanly. On the other hand, as soon as you begin to approach the backyard fence to speak to a neighbor, in goes the left foot into a concealed pile of fresh pooh. It is worse when you are in bare feet. This is the reason why my wife *allowed* me to inherit the job of cleaning the yard of pooh, especially with three large golden retrievers. They are not so diligent either, as on occasion I have seen puppy toe prints in their own pooh. The same can be true when attending an outside wedding and a large bird decides to drop one on your shoulder. *Timing*! Or luck!

There are so many delicate things humans do that require precise timing. The list is enormously long. But again, I need to go back to my family to have you visualize this. Now, it does require very good skill to dock a powerboat amidst strong wind, tidal currents, and other people *telling* you what to do. When I was a young boy, the family would often go out on Pop's boat during the summer, which was docked at our boathouse in North Wildwood, NJ. It was not a very large boat, but we would fish and ride around in the backwaters for hours. However, my grandfather, Pop, drove his boat like he drove his car. The real problem, as you all know, is that boats do not have brakes like cars. So, when it was time to come into the dock, my dad would be up at the bow to grab a line stretched across two pilings, which were used for docking stability. Someone else would secure the boat's stern line and eventually the bow line to the dock. There was a fair amount of timing involved.

Pop would race up to our boathouse and dock it as if he were driving his car up a long driveway. The rest of us would be sitting in the back with white knuckles, gripping the cushions. My dad was up front at the bow, and Pop, driving, would shout out as he entered the docking area, "Jimmy, grab the line. Grab the line. Jimmy, grab the line." My grandmom would be saying, "*Ed, slow down.*" My dad would almost hang himself on the line as Pop powered into the dock area. If boats only had brakes. A quick reverse of the motor, and all was good.

Once secured to the dock, we would take a heavy sigh of relief that we arrived back in one piece. But these were tales that we would laugh about for years. We were definitely entertaining to the neighboring boaters. But it is all in the timing, and our time as a family has so much funny history. Pop, Grandmom, Dad, and Mom are all gone now, but their essence lives in the tales that we continue to tell. *Priceless memories*!

* * *

There comes a time when we all must go our way. The baby bird must eventually leave its nest and attempt to survive. Some animals stay with their parents longer than others. Baby chimps stay with their mothers for up to eight years before becoming independent. Elephants are herding mammals which remain together. When a calf is born, it is protected by the entire herd. They will also come together and return to the site where a member has died to visit the deceased. So much like many human qualities. People can assume why, if they wish.

Some marine mammals refuse to leave their dying mate. A loyal dog may never emotionally get over its deceased owner. And humans are never the same once a loved one passes, regardless of time or age. And the death of your dog is *more* than losing a pet. It is a beloved family member and will always be remembered with everlasting warmth, heartache, and gratitude. We go on only because we must. Take comfort in memories and do not allow them to become sorrows. Think of what you would be like without those tales to tell and tearful memories of total joy. Without them would make us less human.

Time can occasionally be like perception. Both are a part of reality, except for the deceased, where time does not exist. A person experiences differences in the flow of time throughout various stages of life. Occasionally, days appear to endlessly drag on. Other times, they feel like only a few hours, and at other times, the hours may only seem like minutes. And if you are lucky, time stands still at your greatest once in a lifetime memorable event. Those events may feel like moments, days, or even years. It is a gentle touch of heaven which not everyone may experience or recognize.

And from my experience, time slows when news of the death of a loved one touches you. The world all of a sudden appears to be moving in slow motion, at which true time is not relative. Perhaps it is because the perception of time is a quality of human experience. Through our evolutionary travels, we have recognized the passing of time, how quickly life events can appear, and how instantly they may vanish. Time is significant of what humans are capable of becoming. Time, then, is part of our perceptive nature in which our behavior is affected. This enables us to manage it to the best of our ability. We can also lose ourselves in time and may often feel lost because of it. It is human nature's way of being either incredibly gracious or horribly cruel.

We have all felt "lost" in life, either momentarily or for extended periods. I think that it may be part of human nature's design for us to evolve through. Our ancestors would never have been able to defeat their fears, accomplish the risks taken, and develop courage to survive without understanding and using the most of their time. It may have been the first time, some fifty thousand years ago, that early humans began to ponder and plan their lives without direct instinctual reflex responses, with their limited expanded thinking.

People experience their most agony when it is time to let someone go. Human behaviors tend to hold onto emotional attachments to others, as we do our own valuable possessions. And there will be a time for all to let go. We also have the difficult task of releasing ourselves when *death* arrives for us. The trials and errors in life hopefully open our eyes to know and appreciate that it is OK to let go, when it is time.

*　　　*　　　*

There is a unique and interesting concept in the movie **Meet Joe Black**, starring Anthony Hopkins and Brad Pitt. "*Death*," played by Brad Pitt, comes for his intended subject, Anthony Hopkins, only to take temporary human form to "experience" what it is like to be alive as a human rather than the symbol of Death and the taker of lives. Death finds that life is a new experience for him and enjoys many of the limitless nuances that it offers. Never before has Death known what *life* means to the living. His job was to take life when it was time. Pitt's character enjoys each new moment so intently that he finds himself falling in love with Hopkins's daughter in the process. This is a completely new experience of human emotion for him. It is one feeling and desire that people cannot live without. It was a pivotal moment in our human evolution when the feeling or instinct of true love for another changed us to become more human.

Death selfishly decides that not only Hopkins's character but also his daughter should be taken. He just cannot see himself existing without her. It is so true that we often feel as though we cannot continue onward without those left behind or those who travel ahead of us. Death struggles in the end over the idea but reluctantly leaves her behind, understanding that self sacrifice is a quality of love and what *real humans* will do for those we love. Finally, he returns to his realm only with Hopkins, as it was only his time to go. The film shows that even Death learned the value and sacrifice we all make for the sake of someone we love. Pitt's character also understands that to truly love unselfishly sometimes means to let them go, even though it feels like it will destroy your very own being. It is ironic that Death learns about life. The twist in the movie is about our human emotions, which drive us. The nature of it all, however, is that we are to die only when it is time.

What will we miss when time runs out? The idea of being "here" is a human concept that connects us to reality and the living. So where were we before we were "here"? That question may be beyond our understanding. Perhaps we were in a state of mind, waiting to emerge like a bud on a plant. Quite honestly, it does not matter. There is nothing to be determined or figured out. I see no need to explain this unknown just for the absolute knowledge of it. We are not that entitled.

Being "here" for gathering knowledge and continuing our species is one thing. Being "*here*" for loving and sharing our lives with others takes on a whole new meaning. That being said, we have no need to concern ourselves about being "*here*" today. When you are in the hearts of your kin, you never really go. Memories are kept alive by generations within a family by telling tales. So why would we need to seek an absolute truth about something which is most likely indescribable and incomprehensible to the living? Being "here" now is the only important thing. For every breath we take is life. Do not be afraid when time runs out. In my belief, the essence or soul of who I am will continue onward toward a peaceful realm, one that is totally unknown to me now. The energy of life cannot be destroyed.

Humans very often can now accept a belief or unconfirmed speculation about our existence beyond an intangible life, along with those who have moved on. It is a belief or faith edged by evolutionary development and philosophy that has marked our part in this world. It has become human nature to know the unknown. Perhaps it is because, since we are here today, somehow we will always be here tomorrow. We feel this from the footprints left by our past. Our tomorrow is not etched in stone or predetermined. Acceptance of this unknown is confirmed by "*when it is time.*" This phase of life is not dictated by us, and this may be what many people fear the most. This is because humans often think that they are in control of life, until, out of the blue, our lives are taken. You control nothing.

This is the analogy when a leaf is determined to fall off its tree in autumn. It will occur when it is time. When you look at an old oak tree, one understands that it has had many leaves during its life. Each had value, none were wasted. All the leaves must go at some point in the seasonal cycle. For without the leaves, there would be no tree. Without each human being, there would be no human culture and no human nature to try and understand. No human race. Besides, if that were to happen, how could we then possibly laugh at our outrageous human behavior? So, everything is meant the way it is supposed to be, including when it is time. We often have difficulty accepting this because *we are only human.*

I have tried to characterize the factors which make us human. Each day, I find new aspects that qualify who we are. Biological anthropology is an evolving subject. We continue to learn about our past and current changes of events which are almost too rapid to keep up with. To predict future human nature and behavior would be based upon supposition rather than probability. However, we have examined closely that there are still constants which remain within our changing species and past ancestry. Perhaps this says volumes for genetics and predispositions in our genetic makeup.

When you look at a leaf of an oak tree, they all appear similar in structure and design. When you look at individual humans, we see unique behaviors and different expressions of common physical external traits. Yet we are so similar genetically that we are all the same as a species, and only one race, *human.* It leaves little room for discriminating opinions unless one only sees outward appearances, which mean very little. We are all leaves on the *same* tree. Do not waste your time looking for differences. Look at similarities and how we all are uniquely the same, and yet different.

There comes a time in life when occasionally we all must apologize for being ourselves. No one is perfect. The timing, again, is crucial. When we take a long look at all the actions and behaviors among us occurring around the world, one can see so much hope, and yet that which is totally horrifying. When did it all come to this? Our collective history of attitudes, beliefs, nobility, and a hesitant resistance to change has made us who we are. Most people are generally cautious. They may be silently fearful of what is seen, said, or heard as untrue. We continually seek out where the truth lies.

There is no doubt that true evil exists in the world today. Yet most will move on to fight for what is right and good based upon their beliefs. Perhaps we do so because we are only human. Occasionally, we act without thinking, resulting in confusion among others. Justifications for our behaviors may be true or not. For the sake of being wrong, apologies mean little, actions mean *everything*.

If in doubt, apologize to yourself and others for being an ass. It is civil to apologize when insulting others. I get it, but your behavior matters to all. An apology is not a sign of weakness. It is a sign of strength. However, the word "sorry" is said much too often. Are you sorry? Really? When it is truly necessary, an apology is a respectful attempt to make amends. Do not give that up. Say it meaningfully. Remember, insulting actions and words cannot be taken back, and words that are not spoken will also be remembered.

Forgiveness is vital. It may begin with a hard sought absolution to make it right again. Apologies are often given shortly after an offensive incident occurs. There should be no time to wait. But there are occasions when an apology is coordinated at the proper time. Here, the timing of the apology has a greater impact. Time, therefore, plays a role in all dialogue and actions which follow. Everything we do, or do not do, relates to an optimum, precise time for our actions. Did we choose wisely in our decisions, or did we wait too long? Time, again, will tell. There is no secret formula in how to proceed, so good luck with that. They say that time heals all wounds. Unfortunately, no, it does *not.*

Because humans behave so differently under various conditions, there are a few general conditions in how humans find fulfillment. Some people either choose what they want and justify actions for personal benefit or follow others like sheep. Many, fortunately, wish to make a difference in the world. Time well lived will allow you the opportunity to find meaningful purpose.

Another factor includes extrinsic components that are generally outside material influences which offer satisfaction. Yet another is by way of intrinsic aspects, which provide personal reward and self justification. An *extrinsic* component for me includes good coffee in the morning, even the aroma as it brews. Take the time to enjoy its pleasurable scent. And what can we say about chocolate, other than it is a gift? Savor each bite, hide it when you can. There are a few others for me, but thankfully not too many.

People are either *sheep,* followers, natural protectors, *sheepdogs,* or *wolves,* preying upon others. You must decide who you are. Self reflection over time will indicate which type you may be.

* * *

Practically everything else has intrinsic value for me. I love the sight of my family, the sounds of their frequent laughter, their touch, and the way others look into my eyes. In my family, we all respect one another's few wishes that we verbally share. Emotional intrinsic necessities for myself include feeling the water across my body in the surf, the warmth of the sun, and even the cold of winter. Snow, particularly at night, fills me with warmth of the mind and gives peace and comfort.

In addition, the companionship of my golden retrievers includes too many feelings to mention. Their very existence in this world forever remains deep within my heart and soul. It only took me a short amount of time, beginning at a young age, to acquire these feelings about these animals. The affection I have for dogs links me to being human.

I appreciate time, as we all should. The sight and color of the sky and clouds, from early morning to the evening sun, is memorable each day. I sit outside with the kids and dogs, watching and waiting for thunderstorms as they approach. Even my dogs love to stand outside before storms approach to feel the wind rush past our bodies. Their ears flap in the wind, anticipating a change in the weather. If I did not know better, I would swear they were smiling. And the magnificence of lightning in the sky fascinates me. Many ordinary visual experiences are appreciated and tap into past memories, allowing me to reflect on what once was.

I attend every Fourth of July fireworks display, only to be in awe of the colors in the sky and know that I will witness only so many of these events in my lifetime.

The multiple pieces of written and recorded music are as precious and touching as my own visual experiences of previous memories. What music does for my mind, light does for my vision. I can almost see music as I play or listen to it. There are pieces so outstanding that the sound reaches my very soul and will last an eternity. The creative aspect of writing and playing music has a special place within my character. I am fortunate to be blessed by this.

And second to none is the intrinsic value of the indescribable touch of water on the body. You might think that this is physical and extrinsic. However, I am more of an aquatic animal than a terrestrial one. It has more of a psychological impact rather than physical ones at times. Recognizing these and the thousands of other events begins to make me understand what it is to be human. We need to be grateful and observant of the time given to us, regardless of its length in years. So many souls are taken so young. If I could give *anything* to someone, it would be more time.

On the other end of the psychological spectrum, what we do, and our concern for others, is personally uplifting and truly altruistic in human nature. It is seeing the confident faces of people as they learn a new skill. I have been an instructor or trainer in several different fields of academics and intense physical training all of my life. As a teacher or instructor, you never truly know how you will impact another person's life. It is astounding as to how many people we emotionally touch and those who have graced our own presence. Everyone needs a sense of belonging, and the sharing of knowledge and skills given brings us into the collective human group.

People want to be recognized, heard, appreciated, and accepted among our kind. But I think most would like to feel that they have made a difference in the world. That pat on the back, the thank you from others, and those acknowledgments usually do not happen. A true servant will *not* seek or expect these to occur. And that is because part of our nature is to be humble, which tends to allow the instinctual mode from our past to be a part of the group. We seek not to be viewed as an outcast, but as a contributing member.

There is a part of submissiveness in our character which enables us to find our place among the tribe. Often, all we simply need is a little time to prove our worth to others and to oneself. Not everyone wishes to be the outstanding, stand alone, brave one and leader among others. Nature does not allow it anyway. However, the feeling and knowledge of belonging is critical to personal satisfaction in becoming a member of the group.

This becomes apparent, particularly now, with multiple variations of the numerous beliefs held by many that a higher power oversees mankind. We all need to know where we fit in and become accepted by this greater power. It validates how humans feel.

It has taken us over ten thousand years for modern humans to begin to get it right. We are still working on it and hopefully will continue for however long it takes. It will be for nothing if we do not learn and grow from our continued mistakes. Do not be shocked by the many human failures. We have numerous successes. Balance is a human nature leveling device. It is like the tide, which ebbs and flows.

Many different forms of life have hundreds or thousands of different species. We are down to only one. The "*rule*" in life is that once a species becomes extinct, it can never return. We must get it right today and not wait for a miracle to happen tomorrow. Life for us must play out this way because humans must take what we have been given to make it work. It is nature's way.

As we reflect on our lives, we may wonder when or if the sand in the hourglass is up. Our species now has the expectation to find meaning in all that we do and see, but we need time for this. Time for which we innately believe will always be there. Sometimes we believe that which we have accomplished has more worth and value because we have succeeded in the task through mental aptitude. But then again, sometimes it is our nature to prove our dominance over all life, a sad trait.

There are times when we believe that an event has more significance than is perceived. Nonetheless, we will pursue it to its end with the constant search for meaning. Humans create incredible things by using our minds and applying new ideas. We are always searching for anything new, with time to explore it. We may see time as annoying, inhibiting our actions, yet we feel like we will have all the time in the world. In our minds, there will always be one more day. Whatever we begin today, we claim to finish tomorrow.

We do not see time as finite, but in the recesses of our minds, we know it will cease. We rarely appreciate time until it is too late, or when someone is taken from us. But life is all about time and how we use it.

Because of this, humans interpret the world in many different ways. Things may seem logical, but not all events play by the rules of logic or of our needs. Nonetheless, we perceive time as having a beginning and an end. It is most difficult to find meaning under time constraints. Perhaps the search for a meaning to life has the same importance as our arrival and the brief moments that we are here.

Human life enters the world every few seconds, and life is taken at a similar rate. It is this fine balance between chaos and order that allows for the refinement of our species. As they say, "out with the old, in with the new." Those who replace us are to carry the burden of our past and the hope to better our future. Time is the limiting factor in which all of this must play out.

We strive to make things right for all, but to no avail. Life is a struggle with occasional limited tools to succeed. The pace appears fast, and yet also seems to linger on with little result. Do not dwell on time. Enjoy and utilize most of what you have of the time given.

Human nature and human behaviors are forever shifting to adapt to a changing world. As noble as it may be to attempt to figure out life, we should at least live to its potential. As it is stated, "*What man is a man who has not made the world better?*" { **Kingdom of Heaven** }. I believe that this is our ultimate goal in life. For without this philosophy, what kind of people would we be?

When our evolutionary history is continually examined over the next thousand years, maybe we will have more of an understanding and a refining of the qualities that make us who we are. After all, we are only human.

Few are those who hear those unspoken heavenly words in our heads of "*It is time.*" For some, it is a peaceful feeling. For others, it may petrify them. The thought of our departure is unsettling at best. Humans seem to avoid contemplating their own demise, and understandably so. Some refuse to discuss it, and that is OK.

We do not know just what is on the other side of life. Faith or no faith, it is still a mystery that we cannot solve. Humans believe they can solve any difficult task when given enough time. But when it is time, we cannot, and that is most disturbing to many. Perhaps this is why most believe in a higher power which will look after us. After all, our parents looked after us when we were infants and young children. Hopefully, most learn and keep faith within us, and that is a most comforting thought.

Even for myself, when it is time, I will miss absolutely everything, the good and the bad. But this is to say that each single special memory, loving person, event, and perfect piece of music will still come with me. I will never be without them. I am faithful enough to know that the best is yet to come. No more pain, suffering, or anguish over life's challenges.

I will reunite with those who meant the most to me, and the beauty of this is that it will be eternal. In the end, when it is time, it will be clear that everything was *perfect.* When the end comes, hopefully you will recognize that you have experienced a perfect morning, a perfect mate, a perfect family, a perfect sunset, and a perfect life, when it is time.

Ask a million people what it means to be human, and you will receive a million different responses. The passage of time for humans has made us what we have become. We are strong physically and mentally. We are compassionate and self sacrificing, willing to give our own lives to save and protect others. We are family structured, with unprecedented devotion to our young.

We are sometimes reckless, fearless, and always determined. We can be very destructive to our own kind, other life, and our earth. But we try to make amends for our mistakes. We seek absolution for our sins. We are egotistical, sometimes arrogant and demanding, but always expect the most from ourselves and others. We despise excuses, and we *never* give up.

We are intolerant of the cruelty of humans and other living things. We seek out what is determined to be good and what is evil in the world. We worship a Power overseeing humans and life itself. We want and expect answers because we believe we are entitled to know the reasons why. We find it unacceptable, either consciously or subconsciously, that which we cannot control.

We are determined to select our destiny and make choices for ourselves, even among those who live in repressed areas around the world. We choose to have no regrets about how we lived our life. We will die for freedom and honor the dead as if they were still with us. We choose to be independent but know we must live harmoniously together.

We are creative, artistic, imaginative, and innovative. We develop and exploit given skills and abilities. And we are intelligent enough to know that we must pass on our skills and knowledge through teaching and demonstrating our expertise to others by example. The older we get, the more one realizes we are not as smart as we thought when we were young. It is called growth and enlightenment. We understand that it is hard to let go, but that is life. We do all of these things and more, because *we are only human.*

www.ingramcontent.com/pod-product-compliance
Lightning Source LLC
LaVergne TN
LVHW050816061126
851857LV00012B/825
* 9 7 9 8 9 5 1 7 8 4 3 6 0 *